An Arizona Water Primer

Allowing Arizona to Grow Responsibly…

…While Saving the Colorado River

By Norman W. Fain III, PE

Graphic Art

By Dusan Kovacevic

(aka Maelstrom)

Library of Congress Cataloging in Publication Data

Fain III P.E., Norman W.

An Arizona Water Primer: Allowing Arizona to Grow Responsibly While Saving the Colorado River

ISBNs:

Paperback: 978-1-970864-55-

Library of Congress Control Number: 2025927339

First edition.

Website: Anarizonawaterprimer.com

Table of Contents

Dedication

This book is dedicated to all the people who took the time and effort to help me achieve my objective: to provide a media to educate the everyday Arizonan about our water. Over the years, many people have shared with me various approaches they thought would get this message out. Very few of you agreed. There were similarities, but also vast differences. However, all of you were instrumental in helping me to understand a path to accomplish this.

When I put everybody's advice into a bowl and mixed it up, a strategy emerged. Thank you to all the people who offered advice and direction. When taken all together, it became a wonderfully curated strategy where one did not exist before. Thank you all.

Thank you

A huge thank you to my wife, Karen, who put up with the many versions of this over the course of three years. What started out as 3 pages became a primer, in a large part due to her patience and the space she gave me to do this. Thank you!

Thank you to Larry Gaydos and Bruce St. James of "Outspoken" at KTAR radio for the 6th C. You gave me a way to tell my story about where Arizona is growing in Chapter 3. Marrying the past with the present without forcing a fit. Thank you for your insight.

Introduction

Welcome to "An Arizona Water Primer". This Primer is intended to introduce our current water situation in Arizona. I have tried to present this in a way that everyone can begin to understand our water history by describing how we arrived at this situation and how we can effectively contribute to where we need to go in the future. A brief summary of our water history helps us understand that we have been working on solutions for Arizona water supplies for over a century and that we need to continue addressing water solutions well into the future.

This primer is an introduction to water in Arizona, covering what water is, where Arizona water comes from, how it is managed, and why it is needed. Understanding these basics will help us comprehend the limitations of supplies, their implications for our communities and us, and how we can navigate them successfully.

Water makes much of our lives possible in ways most of us might never understand. Perhaps this Primer will introduce some of the benefits and impacts that we have never considered before. I attempt to accomplish this by introducing the state of water in Arizona in a way that is easy to understand, so that we can progress through our growing water needs as a community, working together to solve and resolve our water concerns. This book is a starting point for gaining knowledge and for deeper thinking by everyone.

This primer is relevant because people often become entrenched in their ideas until something or someone introduces them to alternative viewpoints for consideration. Do we have the knowledge to answer these important questions, knowing they will impact the future? Do we grow, or is no growth better? How do we adapt to our changing water supplies? Can we do anything to improve our situation?

If we have never been introduced to the many ideas, we get stuck, and this primer is an opportunity for many to get unstuck as it relates to water in Arizona. The ideas presented in this book are a combination of ongoing efforts and an attempt to envision a potential ultimate water resource solution and what that solution might mean for us all. I offer these thoughts to spark your thinking, as you may have a much better ideas once you grasp the synthesized relationships that water has with everything in our lives.

Throughout this primer, I offer my opinions based on facts and my industry experience. One opinion presented in this primer is that certain areas in Arizona will run out of water in the future without the development of new resources and practices. The positive aspect of this is that we can develop new sources of water and improve our practices regarding water use.

"Development of new resources and practices"

This book works to explain what that might look like. No growth does not solve that conundrum; it just continues it at a slower pace. This thinking fits directly into the social aspects of water.

Throughout history, water has been fought over, and it will continue to be fought over as people struggle to survive. Today, we must work together to develop solutions that will ensure our sustainability. Together, we can generate new ideas and strategies to move forward towards a better life for everyone. Let's avoid the subsequent mass migration out of this part of the world due to limited water supplies and work together to solve our complex water issues. We have advantages that the Hohokam Indians did not have 700 years ago, but we need to tap into those advantages to avoid a similar fate. A space where we can learn from history while using our growing knowledge and access to technology to adapt to our changing conditions.

The value of this is found in the true stewardship of our water resources. This is the only way we can ensure that the future in Arizona, the Southwest United States, and Northwest Mexico will be sustainable, through and maybe despite climate change. Sustainability means that we, as humans, can live in harmony with our changing environment. It means that our food supplies will continue to be a cornerstone of our environment. It means that we can leave a future for those who follow us. It means we can and will adapt as required to achieve these intentions.

I hope that this primer serves as a source of information, helping to establish a foundation of knowledge that enables sound decision-making regarding water in our lives. My goal is to make learning about this subject easy and enjoyable, so that it sticks with you and holds meaning after you've read this book. Ultimately, I want us all to live a better life in the place we choose to reside, without running out of high-quality water.

Hi, my name is Norm Fain. I am a professional engineer specializing in water and wastewater infrastructure, resource development, and management. I have been practicing engineering for over 40 years, and I have had my own consulting firm, Fluid Solutions, since 1998.

I am also a 4th-generation steward of this great state of Arizona. I don't think just for tomorrow; I think for generations to come as well. What are we leaving our children and grandchildren? How will we leave them possibilities for them to succeed? These are important considerations in which water plays a crucial role. We use water in almost every aspect of our lives. To live without it would be impossible. Living with significantly less water would significantly change our lifestyles, and for most of us, reduce the quality of life.

I have created this Primer because, as I listen to people talk about water, I hear fear. Fear of running out of high-quality water or just running out of water, period! These are very real situations that we need to be paying attention to. Running around like a chicken yelling, "The well is going dry," is just as counterproductive as stopping growth. Neither one will help us protect and manage our water supplies. My goal with this primer is to facilitate a general understanding of the state of our water situation, how we arrived at this point, and where we might consider going, so we can act as a community to create and develop ways to improve it for everyone.

When people make decisions grounded only in fear, we often regret the choice. Fear can be a great motivator, but let our actions be grounded in sound thinking. That can only happen when we remove fabricated limitations by understanding the whole situation. Ignoring facts leads to actions that are not in our best interests. I came across a quote by "The Captain @sgrstk" that fits this thinking perfectly:

A fact is information minus emotion
An opinion is information plus experience
Ignorance is an opinion lacking information
Stupidity is an opinion that ignores a fact

My intention with this Primer is to help create a foundation of knowledge grounded in facts, with my opinions offered, as appropriate, based on my experience, to support people's understanding. The purpose is to gain public understanding so they can support sound solutions to our future water needs. When a water deficit is identified, we need to act to alleviate concerns and meet our water needs. Let this Primer help mitigate ignorance and avoid opinions that overlook facts by presenting the water situation comprehensively for your consideration as you work to develop your own understanding of water in Arizona.

Our decisions today will impact generations to come, making these actions extremely important. Do we leave future generations possibilities for opportunities, or do we leave them a disastrous mess? Building on the efforts of the past and continuing to plan for the future is a path to leave our children a bright water future. Our forefathers left us with possibilities for a future; isn’t it our responsibility to leave possibilities for our children? Water has to be part of that thinking.

If you're interested in learning the basics about Arizona water, this primer is a good place to start. It’s a primer, which by definition, according to Merriam-Webster, is a small introductory book on a subject. The book is concise yet comprehensive on water, explaining why it is so crucial to our continued existence in the world as we know it today. Each chapter could become an entire course or even a career if one chooses to dig deeper, but they only scratch the surface in this book.

For some, it may be all you need to quell your fears, enabling you to think and act responsibly. I hope that for others, it serves as the catalyst that sparks your continued learning and growth around water, enabling you to understand and possibly contribute to the solutions. Those who are moved into a water-related career are welcome. It is a rewarding profession where anyone has the opportunity to be a steward of our greatest natural resource: water.

Chapter 1: What is Water

The Most Valuable Resource on Earth

Water is the single most valuable compound on Earth! Think about that, more valuable than gold or diamonds or even any rare earth metal necessary for technology today. Can you live without gold and diamonds? Can you live without water?

Water is required for life as we understand it. When scientists search for life on distant planets, they look for water because life, as humans define it, is biological and biology requires water for survival. When was the last time you went 3 days without some form of water? It feeds our plants, our animals, and our bodies. Without water, nothing biological could live for an extended time. How long varies from minutes to eons, depending upon the biological structure.

Pure water is a colorless, odorless chemical compound consisting of two hydrogen atoms and one oxygen atom to make up the water molecule.

H_2O = HOH = Hydrogen Hydroxide = Water

Figure 1: The Water Molecule

The molecule can be found on Earth in three phases: solid as ice, liquid as water, and gaseous as vapor. When all three phases of water are combined, the amount of water found in the Earth's ecosystem is believed to be finite. It moves in and out of each of the phases, but the amount of total water does not change; we cannot increase the amount, and we cannot decrease the amount (except for the small quantities of gaseous water that escape Earth's gravity into space). That is another point to ponder for a moment. The amount of water on Earth is finite. We only have so much, and quality varies widely between our oceans, rivers, and streams, as well as the differing groundwater sources.

Let's talk about the water molecule itself. Pure water is hungry. Pure water is rare naturally on Earth because it has an ionic imbalance. Because of this imbalance, water tends to bond with other compounds chemically, absorbing constituents such as salts and metals that readily dissolve in the water. One use of pure or near-pure water is part of a cleaning process in the manufacture of silicon chips. Hungry water removes impurities on the chips, effectively cleaning them prior to finalization of the formation. Another, more common use is the dissolution of chemicals that may start in a solid form but easily dissolve in water for use as a liquid.

On the other hand, because salts dissolve so easily into water, they can also overload the ionic balance. When this happens, water stops dissolving the solid materials, and they remain as solids in the solution.

An example to try in your kitchen to understand this better:

- Take your bird feeder and fill it with water.
- Add sugar one tablespoon at a time, stopping to mix thoroughly after each addition.
- As long as the water is still hungry, it will dissolve the sugar.
- When you add that tablespoon, that doesn't want to dissolve but floats around until it settles to the bottom shows the water reached an ionic balance and no longer needed to dissolve other ions into it.

When you mix sugar in water for your bird feeder, if you add too much sugar, you will have sugar water with solid sugar at the bottom. The water was hungry and dissolved the sugar until it got full, then it just left some as solid and did not dissolve it.

As long as water is out of balance, it seeks out substances to dissolve and bond with, or places to release the excess. It is those ions that are positively and negatively charged that we were supposed to learn about when we took a high school chemistry class. Pure water will pick up contaminates from anything that can give up ions and be easily dissolved into solution. The human body, soils, salts, and minerals will all release ions that dissolve into water, which is not in ionic balance.

Water is Bipolar

Water is bipolar!! It destroys by erosion and sedimentation. It takes away, and it gives away without concern for anything but its chemical balance and physical capabilities at a place and time.

Figure 2: Picture of the Grand Canyon from Park Archives

- Water carved the Grand Canyon into the earth. The mighty Colorado River eroded and carried solid silts and sands away. Much of the Grand Canyon's geology contains salts that are readily dissolved as water flows across it. Water dissolved salts and minerals around the rocks, freeing more sand and silt for transport as it carved through thousands of feet of the Earth's history. The faster the river flowed, the easier it became for erosion to occur. As the load of sands and silts in the water exceeds its ability to carry it, usually because the flow slows down, it dumps the heavier materials first, such as the sand, which makes up some of the banks and beaches along the river. Flow slows further, and the silts settle out in the deltas and flood plains around the river.

Such destruction from water can readily be seen in many places on Earth. We readily see this in hurricanes and tsunamis today. Snow melt and rainfall can create floods that destroy environments. Water is corrosive, deadly, and powerful. It appears in many forms, combining its natural destructive forces with winds and other natural weather conditions to greatly increase the ability to cause damage.

- Alternatively, water is vital for biological survival. Humans need it. Animals need it. Plants need it. Without water, we die. Biological beings, as we understand life today, cannot live without some form of water. Just watch your plants wilt if they need more water. Biology, as we understand it today, requires some form of water to be sustainable. The amount varies with the biological being, but the fact remains the same. Water is required to sustain life.
- Chemically pure water will eat away at its surroundings. It will cause the oxidation of steel corrosion. It will attack most materials similarly to a strong acid, typical of a low pH. However, its pH can vary significantly depending on the contaminates.

Some plastics and ceramics can resist the water's corrosive tendencies better than most materials; however, given enough time, water will win the battle with those materials, and they will succumb as well, although some materials may last for centuries before losing to water. The Grand Canyon is a testament to the power of time and water. Water will seep through concrete given enough time. Just look at your local subsurface parking garage in places where water leaks into it. It usually isn't through a crack, but seeps through the wall itself, via the tiny pores filled with calcium and other salts that water attacks to make its path through the wall.

- Water that is overloaded with dissolved salts will deposit salts on surfaces. The ionic balance has exceeded what the water can hold, so it deposits its waste ions. This is the scum on the faucet that sticks like glue and fills in the holes in your faucet aerator. This situation causes things to stop working because they are clogged up with these salty sediments.

Hard water contains ions found in sediments that take the form of dissolved salts, such as calcium, magnesium, and others, when they are in solution. When the water has more of this than it can hold, these salts appear on the surfaces the water deposits them on as they dry. Temperature, evaporation, and velocity of flow all affect the ability of water to dissolve or deposit ionic loads.

Soft water is similar to hard water because it contains salts, but those salts are predominantly sodium. Your water softener may be the type that simply exchanges the calcium salts for sodium salts. You can feel that change on your skin in the shower.

- Aesthetically, drinking water has an ionic balance of salts and minerals to give it flavor and balance. It is neither acidic nor scaling; ideally, it will have a pH of approximately 7.0 to 7.2 units. Adding a chill and "balanced safe drinking water" is refreshing and healthy, regardless of the current source. Did you know that most bottled water providers add salts and minerals back into their water after they purify it? They have to balance the ions in the water, and while doing so, they also either enhance or deter the taste.

Water can physically hurt you or help you survive. That is why the quality is important. Quality in this Primer is all about what is healthy and what is not healthy for biological survival as we understand what that means today.

Water Quality

Water quality is important because water is capable of carrying things that can hurt or help us and our biological counterparts live healthy lives. Water can carry both beneficial and harmful contaminants in two fundamental states. Dissolved into the solution or suspended in the solution. Typically, as a vapor, the water molecule is closest to its purest state most of the time, but there are exceptions when mixed with other gases.

Humans have gotten very good at messing up our water quality with our activities. We have historically used water to create and sustain our inventions and lifestyles—agriculture for our food. Industrial materials, such as plastics, are used in products ranging from computers and semiconductors to tarps and water bottles. Materials for our homes and businesses, as well as parts for our cars. Through this, we have polluted some waters and made them unacceptable for consumption due to our activities in the industrial, agricultural, energy, pharmaceutical, and goods sectors, which are intended to improve our lives. This is not a new phenomenon; humans and other animals have been polluting our waters since the beginning of time.

However, humans have gotten pretty good at cleaning up our messes. In the Middle Ages, when sewage filled the gutters, illness was rampant. When we figured out why, we invented ways to collect and dispose of the waste away from the population. This created cesspools that were undesirable unless they got washed away in a creek or river. When we realized we were polluting the water supplies of our downstream neighbors, we invented ways to clean the water to avoid wars with our neighbors.

The first formal systems for treating our waste were settling tanks of various forms and efficiencies. Then, leach fields were added, and septic systems were put into use. Treatment is limited in these tank solutions, as biologic environments may not be given the time or conditions necessary to facilitate more than a limited treatment to improve quality and remove contaminants. The addition of leach fields incorporated the natural treatment of the earth. To accomplish this, there must be enough vertical distance above the groundwater so that adequate filtration and adsorption are available to clean the water naturally. One significant weakness in this approach is nutrient control removing contaminates such as nitrogen that easily travels through the soils and pollutes the groundwater. This becomes a problem when an entire community is on septic as opposed to a centralized system that can produce higher effluent quality than a septic tank.

Natural systems, such as lagoons and wetlands, became early centralized systems, which shortened treatment from years to months, and were the next common methods employed. These were, and still are, an adaptation of Mother Nature's own treatment methods. Later, mechanical systems came along, which shortened treatment time from months to days. Mechanical systems also took up less space, and odors could be contained and controlled, making them more acceptable in our Cities and Towns.

Through technological advances, we can treat contaminated waters to a very high quality in a short time. Today, we have the knowledge and technology to clean sewage to drinking water quality standards. At the confluence of public acceptance and the cost of producing the water, it will become readily available in many communities. I believe this confluence of cost and need is very near in Arizona. The City of Scottsdale currently has infrastructure that could be converted for direct potable reuse as demand requires.

Part of managing water supplies in arid locations includes conservation in all domains, while utilizing water for its highest and best uses within the community. We also augment this by working to accelerate Mother Nature's ability to purify our water for reuse, rather than taxing the Earth further. In the end, it will take participation at all levels of water use to ensure that we maintain sustainable supplies that support our lives in a manner we choose, rather than being dictated to by our environmental shortcomings. As a community, we all need to work together to build sustainability into our very vital water resources. If any part of a community fails at this, it will make being successful much harder and may even facilitate a catastrophic failure in our water supplies. Personally, I would like to avoid this situation.

We have gone from disease in our gutters to being able to reuse almost all of the water we use today.

And we have gotten pretty good at it…

Consider this…When astronauts live in the International Space Station, they consume water in many ways. In their bodies, experiments, and washing, to name a few uses. Think about the amount of water needed and wastewater generated in a year if the astronauts only used water once before discarding it.

Imagine a big backyard swimming pool for each astronaut. Now, consider a closed system where the water is cleaned and recycled over and over and over again. Instead of storing and wasting large quantities of water, a small quantity is used and reused over and over again to sustain life. That big backyard swimming pool for each astronaut is actually one smaller pool that supports an entire community of astronauts for month after month.

The technology and advancements in water use, treatment, and reuse come from all parts of the globe and many varied sources. These include government, education, business, space travel, agriculture, mining, and manufacturing. Each of these industries needs and uses water, and they all contribute to our advancement in our capabilities to become stewards of our limited water resources.

Chapter 2: Social Water

Whiskey is for Drinking, Water is for Fighting Over

Since the beginning of time, humans have understood that water is required for survival. This is why water is the most valuable compound on earth; we can't live without it. People are aware of this, and when we think we might run out of water, we become scared because we need water for survival.

When a person is faced with fear when death is imminent in front of him, some will freeze, others will flee, but many will fight. Water has often been fought over because biologically, we are programmed for survival.

- Mark Twain coined the phrase…Whiskey is for drinking; Water is for fighting over. I believe that this is one of the truest statements ever made about water. People have killed each other over water in the past, and we will likely continue to fight over it in the future because good, high-quality water is limited on Earth.
- Wars are fought over water. The winners thrive, and the losers die off. This is true whether human or animal in the wild. Whoever controls the water controls their future. Watch Egypt and North Sudan over the coming decades as they grapple over the Nile River and the Aswan Dam.

- The amount of water on Earth is considered finite. It moves and changes form as it fits into the ecosystem of Earth. Current theory is that every drop of water on Earth has been consumed at least once and most of the readily available water many times since the beginning of life on Earth. Mother Nature can clean up any water given enough time, but our human interventions are outgrowing her timetable. Humans have created effective ways to clean up our messes in hours and days, and we must continue to invent new solutions to problems we don't even know about yet.

Now consider these next two thoughts:

- Water is an asset that must be protected and managed.
- Water is a liability if mismanaged.

Reflect on this, water must be protected and managed, or it becomes a liability. Not one or the other, but both protect and manage. No growth may manage water in some places, but it fails to protect groundwater in regions that are not in safe yield. In essence, they are managing the slow demise of their vital water resource as the availability continues to shrink. Communities that are not in safe yield take more water from the ground than Mother Nature and all Human Technology for cleaning and recharging water can produce and replace in the ground over a period of time.

Consider the following formula for the Safe Yield of groundwater for a basin that already mines groundwater:

Groundwater Safe Yield = (All Groundwater Additions/year) > (All Groundwater Withdrawals/year)

This formula simply means that between man and Mother Nature, we must put more water into an aquifer than we take out. How do we do this? In both direct and indirect ways, we can augment supplies.

Water is directly added to the aquifer through recharge. Rainfall that is soft and slow and lasts for several days can put a lot of water back into the aquifer, as the ground can soak it up as it falls. A hard, fast thunderstorm runs water off as surface water that we can collect and store, but much is lost for recharge. We also treat sewage to meet the same quality standards as our groundwater and recharge the aquifers. Typical methods of recharge include surface spreading basins that allow the water to infiltrate, much like rainfall. Vadose zone wells are similar to dry wells in that they do not reach the aquifer in the well. At some level above the groundwater, the well is terminated, and water is allowed to infiltrate the soils and travel to the groundwater. Often used when soils near the surface don't allow percolation to occur fast enough to be effective. A third standard direct recharge method is direct injection wells. These wells are drilled into the groundwater, and treated sewage is injected directly into the groundwater supplies.

Indirect aquifer augmentation is achieved when a groundwater use is replaced by treated sewage or a renewable surface water supply. This indirectly augments the aquifer by leaving water that would have been used or might have flowed away in the ground for future higher and better use.

Water is a fundamental need for human life. It is a requirement for any community to be sustainable and grow. Stopping growth because there is a future assessed shortage of water may not be a responsible solution. That may likely turn a community into a ghost town in one generation after growth is stopped, with few exceptions. Responsible planning can turn the tables on sustainability while allowing a community to experience strong growth. Putting our heads in the ground and hoping that someone will come along and fix things will not get it done. Recognizing that the people and our water uses have outgrown the perceived groundwater supplies, we must work to find new sources of water for tomorrow while continuing to find new ways to conserve and reuse the water we have today. It will take both new sources and conservation to grow into a sustainable future.

Where Do We Get Our Water Knowledge?

Humans have continually found ways to improve our quality of life. If you think about it, the good old days were tough, and average life spans were significantly shorter. We did not have the common conveniences that we take for granted today back then. You do not have to kill the chicken before cooking unless you choose to. You don't have to grow your vegetables when you can get them at the grocery store, unless you choose to. Our clothes are made by others. Doctors, police, and firefighters are available all day, every day.

How we take care of ourselves is changing. Our quality of life has improved and changed faster than ever before due to our past advances. To maintain this, humans must continually explore new ideas and new ways to accomplish things with curiosity and intention. We have to continually adopt new ways of doing things as we adapt to the changing world around us.

The scientific community, comprised of government regulators, academic institutions, and the private sector, including scientists, geologists, and engineers, has been acknowledging Arizona's water situation for over a century.

- Arizona spent from the late 1800s (before statehood, while still a part of the Arizona-New Mexico Territory) through the 1964 Supreme Court Ruling on the Colorado River, fighting for surface waters in our watershed. The Colorado River Compact of 1922 authorized the federal government and the seven states in the watershed to negotiate an agreement for the use of the river water. This was to take effect at the ratification by each of the States that are party to it. Arizona was slow to agree to the Compact, and the

Boulder Canyon Project Act was approved in 1928. It allowed the construction of the Hoover Dam for controlling floods, improving navigation, providing water storage, and generating electricity. It also allowed Parker Dam to serve water to Los Angeles and the Imperial Dam to be constructed, serving the All-American Canal to provide Colorado River water to California's Imperial and Coachella Valleys.

It wasn't until the 1964 Supreme Court ruling in the case of Arizona v. California that some resolution to the states' water rights was reached. This case limited the amount of water California could take from the river and allowed Arizona to keep the water in the Gila River sub-watershed for use in the state. A summary of these cases is provided in the Appendix of the book "Dividing Western Waters," subtitled "Mark Wilmer and Arizona v California" by Jack L. August Jr. For those who want to understand the Colorado River water war, the book is a great read.

- The 1950s through the 1970s brought forth a dawn of awareness about our groundwater at a level that motivated large groups of professionals to assess the deteriorating conditions and begin to institute changes in how we manage our groundwater resources. In part, this situation brought forth the need to establish the Arizona Department of Water Resources to protect people's water rights throughout the state. As a result of the formation in 1980, the responsibilities of the Arizona State Water Commission and State Water Engineer relating to surface water, groundwater, dams, and reservoirs were moved into this single agency, the Arizona Department of Water Resources (ADWR).

- The ADWR was created by the Arizona Groundwater Management Act of 1980. This formally added Title 45 to the Arizona Revised Statutes. This act was controversial at the time, and it still remains so to many today, as they fought for property rights of which water is believed to be a part. However, ADWR has, over time, worked to understand our waters so they can be protected and managed, and that is to everybody's benefit. The Department treats water as an asset, just not always part of a piece of property. Slightly different but significant views.
- The 1980s brought forth the most comprehensive groundwater laws in the United States with the creation of the Arizona Groundwater Management Act. Between 1980 and present day, the codes associated with this law have undergone significant evolution. In accordance with the Groundwater Code ARS 45-401-704 ADWR continues to work on completing its 5th management plan for each individual Active Management Area in the state, with the goal of attaining a safe yield - the state of the water balance

where more water is restored to groundwater than is removed each year or over a fixed period of years.

- Today, the impacted state governments and industries are working on ways to preserve the overallocated Colorado River, which provides water to millions of people and irrigates 1,000s of acres of land growing food, and produces power for 60,000,000 plus people when all is considered. This includes working with the other States and neighboring Mexico in the watershed as well as finding ways to increase the sustainability of Arizona and its people.

These professionals define the pressures surrounding water, including its availability and quality, through the stories they tell. They make these water situations real for the average person. Right or wrong, public perception is borne out by these stories as people adopt different interpretations of what we hear. It is also where curiosity becomes the seed for change and the evolution of new situations that promote growth and a good life for all of us to live with our friends and families.

A space gets created for the people who want to make a difference to come together to resolve new concerns, improving situations for everybody, not just a few. Resolving situations is not saying no, no, no. That does not help the future. It requires foresight and anticipation as well as experimentation and implementation. That means we must listen to all the solutions that are offered because that will allow more great solutions to rise to the surface that we could not see without that collaboration. My grandfather used to say, "I will listen to 100 ideas on a topic because 5 of them I can use." That has always been sound advice that I try to live by because it removes many of my perceived limitations by opening up my thinking to new points of view. The actual solution is rarely a single idea. When all the ideas are mixed together, a new, more powerful approach often develops that could not have existed without the amplification that all the ideas made possible. With that said, the ideas offered in this book can be a starting place for thinking up better ideas. It is definitely not the ending point.

If we let it, all those different perceptions we interpreted from the stories the water professionals created can blend together to form the best answers to approaching and solving our water concerns. We don't have to migrate away because we ran out of water, or stay and suffer because we don't have sufficient water, if we work towards developing sustainable water solutions today for the future.

We have to change the situation to allay our fears around water. What are all the ideas out there that are realistic? What limitations are we willing to work with, and what limitations are we going to work to change? Finally, what limitations can we simply eliminate if we are curious about the possibilities?

Chapter 3: What Does Water Mean to an Arizona Community?

The 5 Cs of Arizona's Attraction (Could there be a 6th C?)

During the pioneer days of the wild southwest, three economic engines drove the economy. Agriculture, mining, and climate. In Arizona, that translated into the five Cs: Citrus, Copper, Cattle, Cotton, and Climate. Each of these industries requires water in different quantities and qualities. They supported growing communities that sprang up around them, giving people jobs and a means to take care of their families.

The communities started with these initial industries, and as time passed, thriving communities have evolved with new economic engines. Agriculture, mining, and climate are still prevalent in Arizona, but other new industries have moved in as a result of growth, technological advances, and human innovation. The engines that drive the communities are changing, allowing the water to be used for homes and businesses where people live and work in higher-paying jobs.

One new economic engine that is getting stronger every year is microchip manufacturing. The 6th C might be "Chips". Microchips, such as semiconductors, and all the supporting appurtenances required to produce them have been a growing industry in Arizona for decades and now support a significant share of Arizona's economic engine. Think of Motorola and Honeywell leading the way, followed by Intel, Microchip, ON Semiconductor, TSMC, and many more supporting industries, with hundreds of companies benefiting from their support. Other technologically advanced opportunities that are moving into the Southwest include data mining and data storage, bringing companies like Apple, Google, and Microsoft into our economies. Each of these new growth industries uses a lot of water to manufacture, energize, cool, and operate what they are providing to the public. These water demands feed comforts we have grown used to that may disappear without sustainable water resources. Your cellular phone would not be possible without water, nor would your TV, microwave oven, and computer. These water uses provide high-paying jobs to people all around the state.

Water is needed to build communities, to help them grow, and to sustain them. I interpret that the wise use of water is water that is always used for its highest and best use in a sustainable way. Water use that is sustainable will help a community to grow and remain vibrant so that it is available for future generations to live a good life without anybody having to leave. Not being sustainable means running out of water. It is a community that uses more water than is restored to a water supply, such as groundwater. To become sustainable, that community would have to reduce its population until a safe yield could be sustained, if it didn't take any other actions to improve its water situation.

It begs the question: Who must leave if no growth is the only option considered? Those who came first, those who came last, the oldest, the youngest? Who gets to choose? Sounds like the start of a war pitting neighbor against neighbor.

The point is that no growth is not a solution to this problem. No growth limits advancements in most communities. The exception includes communities that rely on tourism, and those communities have a seductive draw, such as the Grand Canyon in Arizona, Lake Tahoe in California, Park City in Nevada, or Breckenridge in Colorado. These are the exceptions with an international draw, not the normal situation. Growth is required for the rest of us to avoid becoming the next dying town. It takes a thriving community to sustain a vibrant life in all demographics of the area. Communities that are not thriving have become stagnant, and many of the people who live there are stuck and unhappy. Water makes growth possible. Without it, jobs don't exist, and communities will wither away and die just like a flower in your garden.

Water in a Community Supports Jobs and Lifestyle

A community that thrives will embrace technological change that can help sustain our future and the future of our children and their children. Water is a mandatory requirement for this to be possible. Water makes our comfort features possible so that we can continue to live lives that previous generations did not have. They often could not conceive that the inventions we use in our current lifestyles were even possible, or if conceived, thought of those ideas as space-age fantasy.

Communities in developed countries today require three fundamental essentials; otherwise, people will simply move on, and the community will stagnate or die. These essentials are Water, Power, and access to Technology. A fourth essential is a means to handle sewage safely, but in today's world, sewage is an extension of our water supplies.

Water seems to be the most variable and volatile of the three fundamental essentials. Technology consumes and pollutes water; sewage pollutes and augments water; power generation consumes water as it changes state from liquid to gas; and water supplies are as variable as the weather. If a river flows through the community, it may provide an affordable and sustainable supply that is renewable. However, changing climatic conditions, both seasonally and over long periods of time, can have negative impacts on the sustainability of the resource.

Groundwater is the most common water source in Arizona (and across the United States). But it is a relatively finite resource in the ground. Our aquifers are generally isolated underground reservoirs, and our wells are the straws we use to suck the water out of the ground. Mother Nature typically is slow to replenish groundwater supplies, giving it the "relatively finite" classification because it took thousands of years to originally form. If we suck it out of the ground faster than it can be replenished, we run out of water.

Power is privatized in Arizona, unless individual solar panels are used. Communities typically have a central source of power. Unless there is a river that passes through the community, hydroelectric power is unlikely to be available locally. Suppose solar energy is not utilized on a large scale with batteries. In that case, the likely sources include the conversion of water into steam to turn turbines, which are powered by gas or coal and generate electricity.

Technology is accessible today in most places due to the vast cell phone coverage across North America. Internet service providers follow the users in most cases. In a few places where typical internet service is not available, satellite options are available. In each of these cases, water is heavily used in the manufacturing of our technology tools, from our cell phones and computers to the satellites we have launched into space water was used in some parts of manufacture.

Sewage can easily be individualized with properly designed septic tanks that mitigate pollution to the groundwaters they recharge. However, typical septic systems fail to remove nutrients such as nitrogen, which is a contaminate to groundwater supplies if it exceeds the natural concentration of the groundwater. To keep septic systems from being local polluters they can add treatment to the septic tank or join a regional solution. A regional collection and treatment system is offered by many communities. These are publicly owned and operated, and usually are staffed with cross-trained water and wastewater personnel. Septic tanks are effective, but regional systems usually produce a cleaner effluent that can readily be reused in the community, augmenting existing supplies or recharged directly to the groundwater without contaminating it. Septic systems concentrated in an area have been known to overwhelm and aquifer with nitrogen contamination, common in human waste. This was occurring in Lake Havasu City and the Town of Prescott Valley. In both cases, the Arizona Department of Environmental Quality (ADEQ) reinforced the need to install regional municipal systems in both communities. Municipal systems solve this problem, protecting our water supplies and allowing supplies to be more effectively managed.

Some of the evolving industries in Arizona that are adopting and adapting to change, increasing our water resource capabilities, include:

- Mining. Sand and Gravel mining supports the construction industry, which builds our roads, utilities, homes, and businesses. Rare earth metals such as copper and uranium support our technology and energy industries. Many mining efforts use renewable supplies where possible and recycle the water they do use several times.

  Without these industries, transportation problems would get worse as new roads would not be possible, and older roads could not be repaired. Utilities, roads, and infrastructure would suffer when repairing and upgrading because aggregates and concrete would not be available. Home and business construction would get much more expensive or have to stop because materials would no longer be available locally increasing the associated carbon footprint, adding to the global warming concerns, and potentially reducing the available water even further.

Computers, cell phones, magnets, and other technology would become scarcer and more expensive. Power supplies may not be as reliable as they are today. Water is vital for developing these resources that are vital for sustaining a thriving community.

- Solar power to generate electricity to support the growing community while reducing water demands related to energy. Many energy sources convert water to steam to turn turbines that provide us with electricity. Photovoltaic solar uses no water in the actual production of electricity, thanks to technology; however, it does consume water in the manufacture of the panels and batteries. The savings is found in the recurring water demands that would be used to generate our electricity will be saved in the local environment.
- Regional distribution centers providing jobs with low water use. These distribution centers reduce the carbon footprint

of our goods, indirectly preserving our precious water resources while mitigating global warming.

- Service industries such as restaurants, boutiques, and consultants all use water in different ways to support the community and its vital growth. I enjoy being able to go out to a restaurant or stylist and be confident that the dishware and tools used to serve me have been thoroughly cleaned so as not to spread disease and illness.
- Medical services, including hospitals, nursing homes, and rehab facilities, which rely on water to clean and disinfect the tools that are used to keep us healthy. Without water, these facilities would be incubators for disease, possibly reflecting the kind of care found in a field hospital in the middle of a war.
- Manufacturing of products. Whether it is aluminum manufacturing, foundries, mining, chips, or almost anything else we make, water is required for processes, cleaning, and/or human needs. Water is central to making our lifestyle as good as it is.

As an engineer, I understand that sustainability through growth across generations requires the continual evolution of our use of new technologies by the entire community. We must adopt new ideas to make our water sustainable so we can adapt to the changes in our environment due to growth, climate change, or any other significant change in our communities.

We grow together when we help each other solve our water concerns in ways that do not stifle the community, but allow it to grow in new, inventive ways. Water will always be part of that equation, so we need to work towards its overall sustainability.

The Need to Develop Sustainable Water Resources

Growth in a community requires continual embracing of new technologies by the community as a whole to protect and manage our water supplies. We must all participate to extend our resources and produce sustainable environments to live and thrive. Technological advances have enabled us to use less water to achieve more, ultimately supporting higher wages and a better quality of life for everyone. Think about that, water availability supports higher wages and better lifestyles.

Even our homes have become much more efficient. These advances include low-water-use fixtures and energy-efficient construction, which have reduced per capita consumption across all uses saving our water supplies while reducing our overall costs by using less to accomplish the same thing. According to the Arizona Department of Water Resources, these efforts, in addition to commercial and industrial efforts, have reduced total water consumption across the State by as much as 22% since 1980. In 1980, the state used 9 million acre-feet per year, while in the early

2020s, it stands at roughly 7 million acre-feet per year. When we realize that the state has grown substantially and that reduction includes heavy water use industries of agriculture and energy, in addition to our homes, that number is pretty phenomenal.

We are reaching a point where these conservation methods are insufficient to sustain current growth patterns without the development of new resources and the importation of additional water supplies. This tipping point is brought forth sooner due to the historic drought that Arizona and the southwest US have been experiencing, which has impacted our surface water supplies. That means that we must continue to develop new solutions to our water situation so that Arizona does not experience a great migration to somewhere else. We have the tools to change the situation if we have the will to do it.

Past advances in our water resources have enabled us to protect and manage them, allowing us to weather our hot, dry climate and still grow without running out of water. Growth is the engine that facilitates progress and advancements that allows us all to live a good life. It also brings in the professionals with new ideas that mix with our historic strategies that support and make possible many advances in our abilities to protect and manage our water resources into the future. Growth makes it possible for the service industries to thrive and support our ever-growing needs. Water is required for growth to be possible. Growth is required to sustain our lifestyle.

Let me say this again. Water is required for growth, while growth pays for our water future and sustains our lifestyles.

The economic engines are continually changing. It is essential to think in terms of generations so that the work done today benefits the community tomorrow. When you think about it, we are currently benefiting from the work those before us did. Today, we must work to embrace the future by adopting and adapting to the changes that make lives better. Technology that is helping us to adopt new ways of accomplishing our goals by adapting to its use includes:

- Computers. They allow us to assess information we could not understand before because it would take too long to decipher. We can collect data and review it from multiple lenses with the push of a button at very low cost today. Add in advances in artificial intelligence, and the ability to evaluate and assess situations to solve concerns is greatly enhanced.
- Internet. It allows us to connect to our computers remotely to obtain information without being physically present. Sensors can tell us what is happening in our environment in real time, and the internet allows us to get that information with ease.
- Water and wastewater treatment capabilities have evolved to the level of making direct potable reuse connections possible once the public is willing to accept it. Direct potable reuse,

just like the International Space Station, well, maybe not just like the space station, but a similar intention. Until then, we will recharge the aquifer and withdraw it for potable use at a later time, or reuse it on the golf course, allowing the groundwater to be conserved for a higher and better use. Both are effective, but not as effective as direct potable reuse due to process losses and regulatory requirements throughout the different approaches.

In Arizona, direct potable reuse is ready to become a reality due to a willingness to explore possibilities that technology makes possible. We understand how to do this safely, and the laws will allow it on a case-by-case basis. This is one of the ways the community at large can, and ultimately will, help to extend the use of each gallon of water. People scoff at the statement of "it will happen", but when the cost of water gets prohibitive while the need continues, that direct potable reuse will become acceptable. This is so we can all live here without anyone having to leave, but it is not the final water solution for the Southwest United States and Northwest Mexico. We have to keep pursuing more solutions to our water sustainability as we continue to grow our communities. But I believe it is a required step on our path towards sustainability.

Does this solve all our concerns? No, but it is a big step in the right direction. It is a thoughtful action with the future of the whole community in mind. It does not limit thinking but opens it up to new ideas, allowing us to continue pursuing the wise use of water in our communities and state. Other measures, such as additional conservation and importation, will be required in the future so we can continue to grow and avoid "Ghost Town" status. In some cases, that future is much sooner than people may think.

Figure 3: Abandoned Buildings in a Ghost Town

Chapter 4: Water Resources in Arizona

Arizona currently has four distinct sources of water that can be used to meet the state's growing needs. Groundwater is local to an area. Surface waters flow through an area. Reclaimed waters are generated from waste streams in a community within the same geographic area. Imported water is relocated from one place to another, such as from one groundwater basin to another.

Arizona is one of several of the states that manage groundwater separately from surface water. Statutes define surface water as "The waters of all sources, flowing in streams, canyons, ravines or other natural channels, or in definite underground channels, whether perennial or intermittent, flood or waste or surplus water, and of lakes, ponds, and springs on the surface." Groundwater is subsurface water contained in basins throughout the state. The separation of groundwater and surface water was clarified by the Arizona Supreme Court in 1993. At that time, they defined subflow as waters that slowly find their way through sand and gravel, constituting the bed of the stream. Underground water outside of the immediate subflow of the stream and its banks is not considered connected to surface water.

Groundwater

Groundwater is water sourced from and stored underground. Water is stored in the pore spaces between soil particles below the ground surface. The basin that contains the groundwater is an aquifer. Between the ground surface and the top of the aquifer, the groundwater is transient, moving in the direction gravity takes it, but only as fast and as far as the soils allow it to move. This is called the vadose zone. Water may or may not be present at any given time.

Picture the tree well filled with water; it takes time for that water to seep into the ground. After the puddle is gone, the soil is still wet with water. Sandy soils allow the water to move faster and therefore dry faster, while clayey soils slow the water greatly because the clay has smaller pores. When water is added, clay swells, making the pores even smaller, keeping the soil moist longer. When the water that is slowly seeping into the aquifer gets to the top of the groundwater, it meets a space where the air between pores in the soil has been nearly completely replaced by water.

It is the water stored within the pores below this interface with water and air in the soil that makes up an aquifer. That is to say that the top of the aquifer is the highest elevation where the air in soil pores is replaced by water. Most aquifers are contained within geologic features that isolate them from one another. This results in a situation where aquifer capacity is relatively fixed. As we use groundwater faster than it is replenished, the top of the aquifer will get deeper, making it harder to get water. It is replenished naturally by external sources, such as rain and snow, that percolate slowly through the soil over that specific groundwater basin. We also directly replenish aquifers artificially through recharge activities using both surface water supplies and reclaimed sewage.

Stormwater is the primary source of aquifer restoration in most situations. It falls to the ground, and some of it infiltrates the soil to feed the vegetation. A portion of this continues to percolate through the soils until it finds resistance. As water reaches the top of the aquifer, it refills or adds to our groundwater supplies. When we drill a well into the aquifer that is not under pressure, the top of the aquifer is the level that shows up in the well after it has stabilized. If the groundwater is under pressure, when we drill the well, the water in the well will rise to a level that equalizes its internal pressure against the atmospheric pressure exhibited at the well site. The height that the water rises reflects the amount of pressure the aquifer is under. This is the difference between confined and unconfined aquifers; one is under pressure, and the other is not. Artisan springs are often sources from confined aquifers that have pushed the water up through fissures in the soil.

In Arizona, several types of aquifers containing groundwater. A homogeneous aquifer is one that has consistent soil formations throughout. It can be filled with homogeneous mixes of sands, silts, and cobbles. The reality is that most aquifers are not homogeneous but contain variations in soil structure because they are formed by varied geological and hydrological events over great periods of time. Add some clay hills and stratified lenses of clay to further confine the space where the sands, silts, and cobbles can occupy. When a well is put into the sandy-silty aquifer, where water is plentiful, the pumping capacity is likely high. When the well is in the clayey aquifer, where the water is minimal, pumping rates will typically be low, if any water can be extracted.

A variant of this aquifer features clay layers that overlay the aquifer. When the source of recharge is laterally through groundwater flow or movement, pressure can increase in an aquifer as it is recharged from sources not impacted by the clay. This is a confined aquifer. When a well is drilled into this water-bearing aquifer, the water surface in the well is above the actual top of the aquifer. The rise in the well will continue until the pressure equalizes with the atmosphere. The height of this water surface above the top of the actual aquifer surface reflects the amount of pressure the aquifer is under. Artesian wells and springs are sourced from this type of groundwater as the pressure is sufficient to push water above the ground surface through cracks and fissures in the soil matrix.

Another common aquifer found in Arizona is fractured. The water is found in fissures and fractures in rock formations. This is a much more difficult aquifer to manage because the water can be much harder to find. It will contain much less water per cubic foot than other systems because the volume of pore spaces that can hold the water is much less. Each space may be larger, but the total volume of space will be less because the number of spaces is greatly reduced as it is replaced by solid rock formations. As I have heard my grandparents say many times, "you can't squeeze water out of a

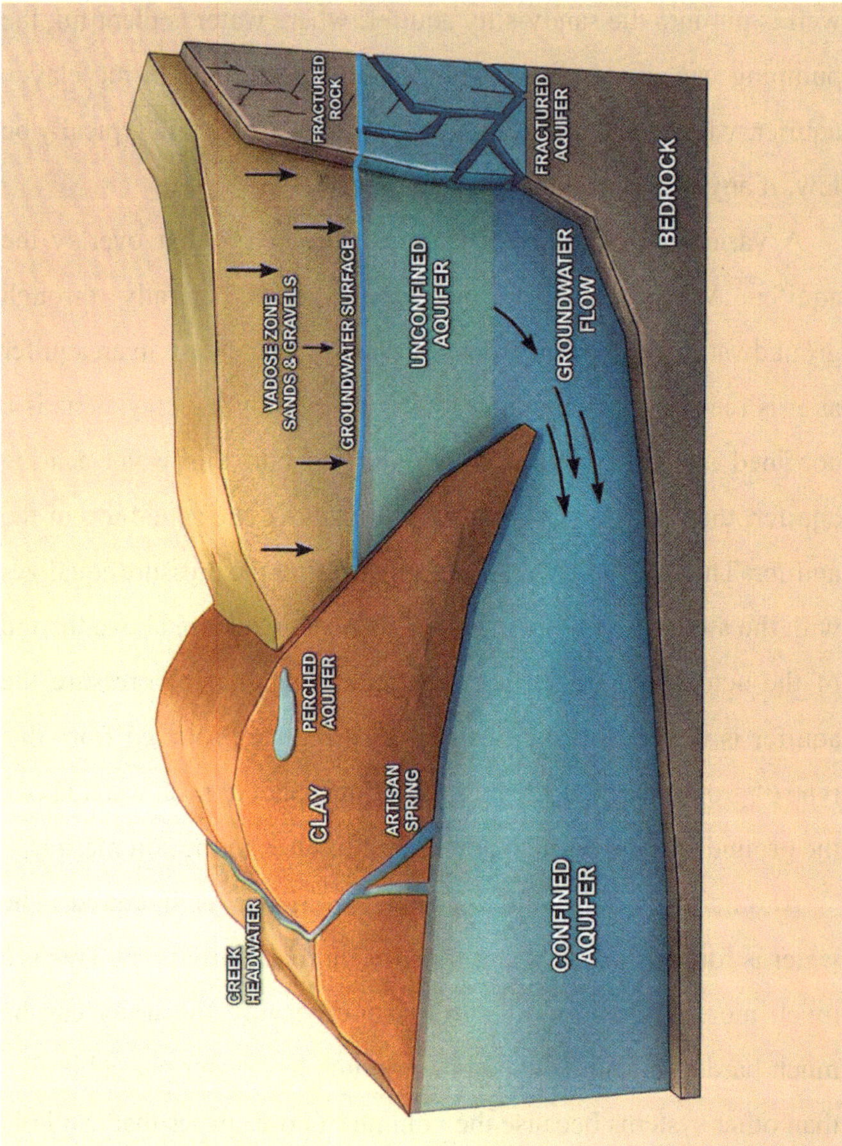

Figure 4: Common Aquifer Configurations

rock". A metaphor used in many situations where things will not work, but is extremely true when taken literally.

Sometimes a perched aquifer exists. They are usually very small, often of poor quality, and not usually sustainable. They exist due to isolating conditions in the soil matrix that hold the water in place above the true aquifer. They appear as a small pond in the profile of the strata, located in the vadose zone.

These are just some of the aquifer conditions that can be found in Arizona. This is just an introduction to our aquifers; there is much more to understand when considering each local situation and the use of the water. No two aquifers will be the same, and no aquifer will be completely homogeneous. Two important things to understand are that water only exists in the pore spaces and aquifer sustainability is limited due to their fixed capacities, as refilling them takes more time than it does to pump the water out of the aquifer. They recover very slowly because of these phenomena. Our aquifers are not bathtubs, but people often think of them that way and it is an often-incorrect analogy for any given aquifer, but it is a convenient way to explain the confining conditions that create the limits of the aquifer.

Finding groundwater is one aspect of groundwater in Arizona. The right to use it is an entirely different perspective to view Arizona water through. The Arizona Department of Water Resources (ADWR) governs groundwater in two different ways, depending on whether you are in a designated Active Management Area (AMA) or not. If inside an AMA, water is governed as if it were an asset held by the state for the benefit of the people. Individuals do not own the water; they hold a right to use the water through a permitting system. In Arizona, groundwater rights are a function of where in the state they are looking for water. Inside an Active Management Area (AMA), water rights are tightly controlled and monitored. Outside and AMA, water rights are tied to the ability to get it with limited governmental control, actually supporting the concept that water is owned by the property owner. This is more akin to having the water as an asset that is part of the land. If you have the means to drill a deep well and produce 1000s of gallons per minute outside of an AMA, nobody will stop you because you have the right with the land you own.

AMAs are areas that have been identified as having water sustainability concerns. Aquifers that have been identified as not in Safe Yield are one of the primary reasons for the formation of the AMA. A primary goal of the AMA is to find ways to attain a Safe Yield condition. Methods include conservation tactics, reuse and recharge efforts, and importation from other sources of water. As ADWR continues to increase its understanding of our aquifers, new designations of areas requiring additional protection and management are developing. New information about our aquifers is also forthcoming that could change how we think about them today. This is done with the intention to help people live a good life without hurting our neighbors. Arizona currently has seven AMAs and three Irrigation Non-Expansion Areas (INAs) limiting the addition of more agriculture. The four initial AMAs include the Prescott

AMA, Pheonix AMA, Pinal AMA, and Tucson AMA. The Prescott, Phoenix, and Pinal AMAs have the goal of safe yield. The Pinal AMA has a depletion criterion.

The state continues to add AMAs to the list as they are able to better define groundwater conditions. Added AMAs include Wilcox, Santa Cruz, and Douglas AMAs.

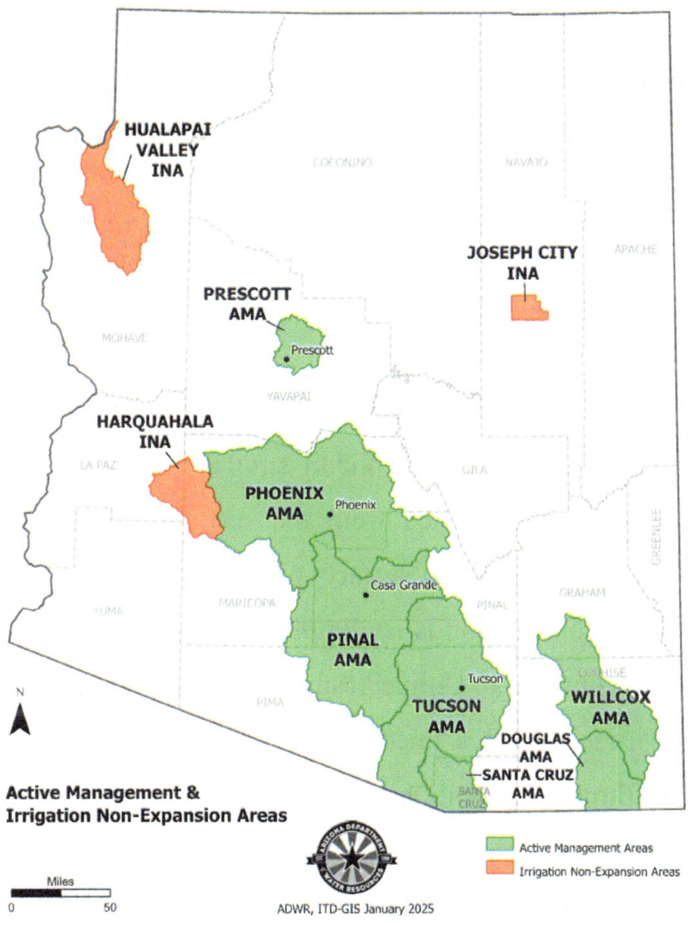

Figure 5: Groundwater Management Areas

Surface Water

Surface water is water that recurs at regular intervals from the atmospheric vapors that form as rain and snow, becoming surface water until it infiltrates the ground. Surface water is a body of water on the surface of the ground, including the ocean, lakes, rivers, streams, creeks, and ponds. Surface waters and groundwater typically interact with each other, feeding into and influencing one another. Lakes, rivers, and streams feed our aquifers, and some artesian aquifers feed our lakes, rivers, and streams. In some cases, the groundwater serves as a base flow, interacting dynamically with surface flows below the ground surface. This is believed to be what is happening along the lower Colorado River, where the difference between the river's base flow and the aquifer has been in dispute in places.

As a side note, the canal systems in Arizona are typically considered as a conveyance conduit similar to a pipe. Therefore, canals have been exempted from the surface water rules, even though they often convey surface water. If they are lined, any groundwater contribution is negligible until the efficiencies of the end use are considered. Unlined canals will allow some infiltration, but even these are often lined with compacted clays, which have limited infiltration capabilities.

Arizona has both internal and external sources of surface water. Internal sources are watersheds that originate within the state's boundaries. External sources of surface water include watersheds that drain across more than one state or nation, such as those in Arizona.

Water that is generated within our state includes some large watersheds that feed significant rivers. These rivers have headwaters that start in Arizona. Significant examples include the Little Colorado River, Hassayampa River, Verde River, Salt River, Bill Williams River, Santa Maria River, Big Sandy River, and Agua Fria River, to name the major contributors. These rivers are supported by smaller rivers, creeks, and washes that support the river system, such as the San Francisco River, Tonto Creek, Burro Creek, Sycamore Creek, Oak Creek, and Lynx Creek.

Just as the rivers within Arizona have sub-watersheds, much of Arizona is considered a sub-watershed of the Gila River, which in turn is a sub-watershed of the Colorado River. Both the Gila River and the Colorado River have headwaters in other states. The Gila begins in New Mexico, while the Colorado begins in Wyoming. This part of the Western United States can be considered one vast watershed, spanning from Wyoming into Mexico, that collects water from the slopes of the Rocky Mountains to the Sea of Cortez.

Arizona has two major watersheds that flow across our international border with Mexico. The mighty Colorado River flows from the northern reaches of the USA into Sonora, Mexico. Historically, it has reached the Sea of Cortez, but a combination of drought and current water allocations along the river has hindered

Figure 6: Arizona's River System

that ability. The second significant watershed is the San Pedro River, which flows north from Sonora, Mexico, into central Arizona, where it is a tributary to the Gila River and subsequently to the Colorado River.

Like groundwater, surface water is also controlled by rights. Before the 1919 Public Water Code, rights were obtained on a first-use, first-priority basis in accordance with Federal laws. This is a "first in time, first in right" philosophy. Through the Public Water Code, Arizona added a requirement to all subsequent water right filings that required continuous use with no gaps longer than 5 years in order to maintain the right. Considerations before 1919 are currently prioritized by the date of first use. Today, considerations after 1919 require the continuous use clause to maintain the right. An old right with a high priority can keep a new right from being available in times of drought and shortage. Water rights and water law can get very complicated very quickly and usually require the assistance of a knowledgeable water attorney who is looking out for your interests. I know first-hand how emotional water rights can be as we fight for sustainability. I also know that if we choose, we can find a path through that leaves everybody better off.

Our surface water supplies are complicated. Who has the right to use the water? Who has priority rights when there is not enough water to go around? How is this monitored and controlled? River adjudication in Arizona is ongoing and will help determine who has priority rights and who has no rights from a legal perspective. Adjudication is a formal court judgement on water rights and is slowly moving through our surface water systems. It is a judicial process that determines everyone's water rights within a river system of watersheds. Since the Gila River is the biggest sub-watershed in

Arizona and encompasses much of the state, the logical movement of the adjudication process is from the top of the watershed at our border with New Mexico and progresses to the bottom near Yuma, where it merges with the Colorado River. In Arizona, that starts at the Arizona-New Mexico border on the Gila River. The process is moving west across the state, picking up each sub-watershed in a logical progression of the water's flow throughout the state.

This will become more complicated as people and communities position themselves to survive the court process without losing access to precious water. Legal adjudication is an attempt to balance state, tribal, federal, municipal, agricultural, and private claims into a court-recognized framework that can be enforced. Ultimately, it will redistribute the available water in a community and disregard any excess rights that cannot be supported by long-term historical uses in accordance with the prior appropriation doctrine that has been adopted by the state.

The only way to protect surface water rights in the community until the higher use needs them is to use that water beneficially, so the courts cannot arbitrarily take the water away. Additionally, the only rights that can be used must have priority over other users in the watershed. This is defined by the date of the right, and complicated by whether the right is pre- or post-1919 Public Water Code. The file copy of the code can be found in the Arizona Memory Project under the water codes of the State of Arizona. It was approved in the regular session of the fourth legislature of the State of Arizona. A summary can be found on the azwater.gov website under the "Surface Water" definition and also defined by ARS 45-141.

Reclaimed Water

Reclaimed water is water that has been previously used, cleaned, and made ready to reuse again and again by man-made means. That sentence says a lot, so let's dissect it.

Water, as we know it, can have various levels of cleanliness. Previously used. Hmmm. Hasn't all the water on Earth been previously used? Current theory is that since life existed on Earth, all water has been consumed at least once. Therefore, technically, all water has been recycled and reclaimed for reuse, either through natural processes or accelerated by man-made systems and processes.

Cleaned and made ready to use again. How will it be used? That defines how it must be cleaned. To irrigate crops while reducing the potential for groundwater pollution with fertilizer, reclaimed water that has not been treated for nitrogen is an effective fertilizer that reduces costs and, due to plant uptake, protects the environment when applied properly. However, that same water may be unsafe to drink due to high levels of nitrogen. But when applied to the crops appropriately, the nitrogen is consumed along with the water by the vegetation. The groundwater is protected from nitrogen contamination, and water supplies are augmented by reclaimed water when used for irrigation instead of groundwater. Finally, the farmer spends less on fertilizer and its application, helping to keep food prices in check, just one of the many ways water impacts our lives.

To recharge the aquifer in Arizona, the quality must meet aquifer water quality standards, which in many cases will meet drinking water quality standards, because most of our aquifers are safe to drink as they exist in nature. Keep in mind that safe drinking water may not meet your taste expectations. Quality for safety and aesthetic quality are two separate considerations. Under current regulations, quality for safety in drinking water is a primary standard and is enforced rigorously across the US. Aesthetic quality is a secondary standard that is recommended but not enforced unless done so locally.

When we reclaim sewage, the end use defines the quality of reclaimed water we produce. There are many variances that illustrate the diversity of water qualities that could best serve a community.

Using reclaiming water, such as sewage, accomplishes several things. First, the contaminates in the sewage will not be spread throughout the environment because appropriate treatment has been applied to remove them. Virus and pathogen transmission is greatly reduced if not eliminated by containing and destroying them in a controlled environment, such as a regional wastewater treatment plant.

Second, reusing water reduces the demand on other water supplies, such as our limited groundwater resources. By using treated effluent on a golf course, the potable groundwater that would have been used is saved, while lower-quality, less valuable reclaimed water is used. This is referred to as in-lieu aquifer augmentation. It replaces the high-quality groundwater supply, so it is not used for that purpose, leaving it in the ground, saved for a higher and better use, such as potable water sources.

Third, when recharged directly into the groundwater system, the drain on the aquifer for community water demands is reduced or even eliminated. This directly accelerates what Mother Nature takes decades to accomplish, making more water available to sustain a thriving community.

Reclaimed water is the result of our efforts to extend our water supplies. Instead of just wasting our "dirty water", we are choosing to clean it up using our technological advances. In doing this, we reduce the time Mother Nature takes to clean our water and make it available for our use in a much shorter time.

Water rights for reclaimed water are often attributed to the entity that developed it. In most cases, the water supply is collected by a central system owned and operated by a government or quasi-government entity. However, this is often paid for initially by a developer and ultimately by the end user through capital purchases and ongoing water consumption. Today, it is typical for developers to consider the effluent produced by their early users as a result of their development as part of their efforts to obtain 100-year assured water supplies. The water generated by the development is treated and recharged to the aquifer to offset future groundwater withdrawals by the development. Development pays for the infrastructure to accomplish this, and gets its return from growth as people purchase the homes and build businesses that use the water.

These actions help extend our water supplies for our use, allowing us to live a good life. It is tactics such as those described above that have helped us to maintain a 100-year assured water supply when planning for growth. These have allowed us to extend the sustainability of our finite resources, literally increasing the amount of water available to a community to use. It has been key to reducing the amount of water used in the state while experiencing robust growth over the last half-century. These actions have been key to our ability to meet our growing water demands over this time. Still, that superpower is running out of possibilities to continue for the next century. Therefore, we need to explore new sources.

Imported Water

The Pacific Ocean or the Sea of Cortez (Gulf of California)? A future possibility. Sea water treated for use in coastal communities and farming in exchange for Colorado River water? It is an ongoing discussion today between states and countries. The scenario could take several different twists between now and its reality. But the process is continuing in Arizona with continued investment over the coming years and decades to help assess how to import more water and preserve what we have in Arizona through new practices in how we use water.

One scenario that has been discussed is the construction of an ocean desalination plant off the shore of a coastal California city in exchange for their surface water allocation in the Colorado River. This may be the most cost-effective solution from a purely capital cost perspective today, but it may not be the most reliable solution. If the Colorado River watershed continues to produce less water, even those rights may not have sufficient priority to retain water in Arizona. As of today, this evaluation is in its infancy. The question is not can we do it, but when will we be ready to import ocean water, politically, environmentally, and financially?

Another scenario would include the construction of a seawater desalination plant and a pipeline into southern Arizona. This may be the costliest, but it may also be the most reliable and sustainable solution. This is being considered from Southern California or from Baja California. Either solution would provide a sustainable supply of water, but a mix of both could make water plentiful through changing climate and weather conditions, potentially for the next 100 years and beyond.

- The Southern California alternative has a couple of twists and options that are possible.
    - Construct a desalination plant for the Metropolitan Water District of Southern California, serving Los

Angeles and San Diego, in exchange for their Colorado River water rights. This would move 500,000 ac-ft/yr of water that is being taken out of the watershed back into the watershed for reallocation. These cities would get a water supply that is not connected to the drought in the Rocky Mountain range through seawater desalination in exchange for giving Arizona its high-priority Colorado River water.

- Construct a second desalination plant off the Los Angeles coast. One would replace the regions' Colorado River water, while the other could use the existing infrastructure alignment to pump treated ocean water into Lake Havasu. This reuses portions of the existing multi-barrel aqueduct/canal system that feed water from Lake Havasu to Los Angeles by reversing the manner in which they operate. This solution actually augments the river flows by adding quantity from outside the watershed into the watershed. It has a double benefit that could add 1,000,000 ac-ft/year to Arizona's water supply, with half of it not reliant on river flows as it is directly imported from the Pacific Ocean.

- The Baja California alternative also offers some unique possibilities that could benefit both Arizona and Mexico.

Currently, Mexico has an allocation from the watershed, but the river has not always delivered enough water to support that with a water quality suitable for growing crops. The US has had to desalinate the Colorado River south of Yuma to meet the terms of the agreement with Mexico because of the natural salts the river has collected over the course of its travels across the US. Much of what does get to Mexico is very salty from agricultural activities and geologic conditions upstream. (The Salt River is named that for a reason; it dissolves natural salts at its source.) This is partially caused by less water in the main stem of the river to dilute salty tailwater or Salt River water. A desalination plant off the coast of the Gulf could restore Mexico's water while also adding supplies to either the All-American Canal and/or Lake Havasu for distribution in California and Arizona. Treat the water to Colorado River quality and deliver water to Mexico while also conveying water to Arizona and potentially California. Mexico's garden belt could be revitalized, enhancing its economy and job opportunities. Routes could follow existing roadways, mitigating environmental concerns. If all of the All-American Canal diversion were replaced with water from the Sea of Cortez, over 3,100,000 ac-ft of water could be kept in the river system. Initially, this water could be used to restore Lake Mead and Lake Powell, and keep our lights on as electricity is continually being produced through the dams.

Combining both of these solutions could correct a significant error made when allowing early water to be transported out of the watershed. Hydrologists warned against this at the time, citing sustainability concerns. Turns out they were correct. Moving water to users in areas that are outside the watershed to treated seawater corrects this historic mistake. However, getting this done will take decades to complete, along with the will to get it done by many diverse parties.

Interim scenarios might include moving water within the State from a very large designated aquifers with mitigated growth; therefore, it has the aquifer capacity to support current and future growth by pumping into a needy neighbor, such as the Cities of Buckeye or Queen Creek in the Phoenix management area, or Prescott, Prescott Valley, and Chino Valley in the Prescott management area. This is an internal importation from one aquifer to another. To accomplish this, Arizona has designated some aquifers that are not expected to serve significant populations and have significant water resource volumes for this purpose.

What does all this mean? One thing it means is that regulatory, scientific, geologic, and engineering communities recognize that we have an ongoing water problem that we are working to solve. We recognize that we will have to import water if people continue to want to live where there is insufficient naturally occurring water, and growth is expected to continue. We understand that growth is required to keep a community vibrant and sustainable, and without it, history has shown in most cases that the community becomes a ghost town with empty buildings and increasing crime.

Through the importation of water and sound water management within the state, there are very few limitations that we cannot overcome. Humans are pretty good at cleaning up our messes. To survive without being forced into leaving, possibly as the Hohokam Indians were over 500 to 700 years ago, we must continue to improve our water situation. Water covers much of the earth, and our oceans consist of most of this water. It is reasonable to assume, especially considering rising ocean levels, that this solution might be around supporting our needs for a millennium or more.

Imported water will create a whole new class of water rights. Considerations for ownership might include those who paid the capital costs or guaranteed the repayment with their hard-earned assets. Farmers have done this in the past and got the Central Arizona Canal (CAP) constructed, along with several dams in Arizona that we all benefit from. In the future, this cost will need to be borne by a much broader set of beneficiaries. Need is another consideration. In some cases, ownership will be those who built the infrastructure, and in others, it will be those in need because they are already facing water shortages. In all cases, there will be a high cost that, if spread across all beneficiaries, becomes affordable due to growth.

Chapter 5: Water Management in the Colorado River Basin

How is the Colorado River Basin Water is Organized

The Colorado River and its tributaries flow through 7 western American states and 2 states in Mexico. The entire watershed includes portions of Wyoming, Colorado, Utah, New Mexico, Nevada, Arizona, California, and finally Sonora and Baja Mexico, before it dumps into the Sea of Cortez. Prior to any formal management of the river, claims in the US were being made in accordance with Federal water laws, with the first in order having the highest priority. Since southern California developed earlier than many parts of Arizona and Nevada in the lower basin and all of the upper basin of the watershed, it acquired some of the oldest water rights on the river. This gives them some of the highest priority water, meaning; when there are shortages, they are the last to have to ration due to the date of first use being before other users throughout the river's watershed.

Some of the oldest surface water rights in the lower basin are held by farmers in the Imperial and Coachella Valleys of California, as well as in the Parker to Yuma reach of Arizona. These rights along the Colorado, as cited in Dividing Western Waters by Jack L. August Jr., make up the largest fraction of private ownership of water rights on the river. Most of this water is moved outside of the watershed through the All-American Canal, constructed at the Imperial Dam, just north of Mexico. Diversions through the All-American Canal account for approximately 3,100,000 ac-ft/yr of the river water usage that is no longer available within the watershed.

Tribes and Cities make up the largest share of water right ownership that is not privately held in the lower basin, with the Metropolitan Water District of Southern California holding the single largest municipal right at approximately 500,000 ac-ft/yr, which is currently distributed between San Diego and Los Angeles. Most of the tribal ownership of water rights is within the watershed, but Los Angeles receives water that is pumped from Lake Havasu out of the watershed. The Parker Dam was constructed to create the initial source pumping station for the Colorado River Aqueduct.

I mention the fact that between diversions at Parker Dam and diversions at Imperial Dam, approximately 3.6 million ac-ft of water is moved out of the watershed. At the time of the early distribution of water allocations within the watershed, hydrologists stated that they believed that moving water out of the watershed would eventually be unsustainable. That appears to be prophetic now. So, what do we do about it? How can we adapt?

In 1921, Congress passed an act authorizing seven states and the federal government to negotiate an agreement around the waters of the Colorado River. In 1922, this agreement, the "Colorado River Compact," was signed by the parties. The agreement apportioned

7.5 million acre-feet of water to the upper basin and 7.5 million acre-feet to the lower basin. The natural divide between the basins is marked by Lee Ferry, located in what is now part of Lake Powell. The purpose of this agreement was to preserve water for states that were developing at a much slower rate than California. Each of the other stakeholders wanted to ensure water would be available once their growth needed it.

The United States and Mexico were also negotiating Colorado River water rights. In 1944, the two countries agreed that Mexico would receive an allocation of 1.5 million acre-feet per year. That changed the total expectation from the river system to 16.5 million acre-ft per year.

Governance of the Lower Colorado River Basin

The Federal Government has generally managed the Colorado River watershed with input from the seven participating States. This included the assessment of the available water in the river system and how distribution has been prioritized. The intentions are noble, but the watershed is grossly overallocated in the current long-term weather cycle, which has been in drought roughly since the 2000s began. Overallocation was predicted by hydrologists when the Federal Government created a standard that allowed water to be removed from the watershed for beneficial uses, such as much of the Imperial and Coachella Valleys and cities of Los Angeles and San Diego.

The government did not heed the warnings of hydrologists who cautioned against using the water outside the watershed, citing sustainability concerns. Approximately 3.6 million ac-ft of the California water allocation is outside the watershed, including both Cities and agriculture. Without the uses outside the watershed, the allocated demands on the river would be reduced to approximately 12.9 million ac-ft per year. Approximately 21.8% of the current overall allocation would not have been used over the years, meaning that the river could have possibly sustained this drought if uses had been kept within the watershed. But that ship has sailed, and whining about it accomplished nothing. So, what do we do to adapt? Can we change it in a way that everybody is a winner, maintaining food, electricity, and water?

Figure 7 generally shows each of the places outside the watershed that are served by the river. It also shows that Arizona is the largest stakeholder by land mass in the watershed. My efforts in this book are focused on the Lower Basin since most of Arizona is in this basin, but some parts of this book may help the Upper Basin as well if they can find ways to develop new water resources or participate in the solution to gain wet water back from the river through importation.

Figure 7: Colorado River Watershed with Out-of-Watershed Uses

The Colorado River Compact of 1922 assigned the duty to distribute the rights to the states. That meant that in the lower basin, Arizona, Nevada, and California had to come to an agreement amongst themselves to distribute the 7.5-million-acre-feet/year allotted by the federal government.

The underlying purpose of this compact was to ensure that the states that developed more slowly would be protected. It was driven by concerns over California's efforts to gain as many priority rights as possible. This action by California, along with the need to secure Arizona's water future, prevented Arizona from ratifying the Compact. Arizona wanted a specific amount of the mainstream water for future use because the state is developing behind California. Arizona also wanted exclusive rights to the Gila River sub-watershed and the excess water the river occasionally produces in a given year.

During Arizona's time of discontent with the federal apportionment of the surface waters flowing through it, California and the US government proceeded to build Parker Dam with the purpose of diverting water to the Metropolitan Water District of Los Angeles. They had already constructed the Imperial Dam to divert flows into the All-American Canal to irrigate Imperial and Coachella Valleys. During this period, the State Department completed negotiations with Mexican counterparts to deliver 1.5 million acre-feet of water to Mexico. This was the Mexican Treaty of 1944, which added to the hypothetical water available, increasing arbitrary allocations from 15 million acre-feet per year to 16.5 million acre-feet per year. The 16.5 million acre-ft that the government allocated from the main stem of the Colorado River are summarized as follows:

Portion of the River System	Main Stem Allocation
Upper Colorado Basin	7,500,000 ac-ft/yr
Lower Colorado Basin	7,500,000 ac-ft/yr
Mexico	1,500,000 ac-ft/yr
Total Main Stem Allocation	16,500,000 ac-ft/yr

When we break down the 7.5 M ac-ft/yr allocated to the main stem of the Lower Colorado Basin, and we consider the amount of the water that is shipped out of the watershed as it creates an interesting picture. If a source close to the users outside of the watershed could be made available, what could that do to the river system?

The lower basin was ultimately allocated as follows:

Stakeholder	Distribution within Watershed (ac-ft/yr)	Distribution outside Watershed (ac-ft/yr)	Total Allocation (ac-ft/yr)
Nevada	300,000	0	300,000
Arizona	2,800,000	0	2,800,000
California	800,000	3,600,000	4,400,000

If we could develop a sustainable way to replace 3.6 MAF/yr for California cities and farmers with a reliable and sustainable water resource, everybody's water situation would improve. The

Colorado River could be saved and allowed to sustainably serve the watershed that gave it birth. Arizona has never been satisfied with this situation, and we have battled California in court over this very issue.

Arizona subsequently sued California to mitigate the damage and gain some control over its water future. Ideally, Arizona wanted the river to be limited to the watershed, but that battle was a losing effort. Arizona also fought for a bigger share of the excess water during the wet years and wanted the waters of the Gila River exclusively. In 1964, in Arizona v. California, the US Supreme Court issued a mixed ruling. It allowed California to retain its existing water rights outside the watershed, but also granted Arizona exclusive rights to the Gila River and an equal division of excess water with California. This had an indirect effect of reducing the water into the main stem of the Colorado by all the flows from the Gila, making the arbitrary allocation of the main stem even tighter. However, it gave Arizona control over how the Gila River sub-watershed was managed and used in Arizona. The control that Arizona gained has yielded very different results between the two approaches to watershed management during this period of drought.

Differences Between Gila River and Colorado River Management

Today, the allocations on the river have been proven to be more than the mighty Colorado could deliver during this current multi-decade drought. Lake Mead and Lake Powell are both dancing with

30% capacity or less as of 2025. When full, either of these lakes holds more water than all five of the upstream reservoirs combined on the river watershed. These two lakes are two of the largest man-made reservoirs in the US. When full, the US Bureau of Reclamation (USBR) cites Lake Powell as having a capacity of 24.32 million acre-feet (MAF) of water, and Lake Mead has a capacity of 28.13 MAF. Combined capacity is 52.45 MAF, while the five upper reservoirs have a total combined capacity of 34 MAF.

Due to the drought causing the lakes' reduced capacities, the Lower basin is facing shortages resulting in a reduction that has amounted to a significant portion of Arizona's Colorado River allocation and will likely climb if the drought continues. The federal government wants an across-the-board reduction in water usage and has introduced multiple paths. The first path is that the states solve the problem themselves. Second, water is cut by priority, which could leave California with the only water. The third is an equal cut to each state by percentage of their rights. This is cited with the US Department of the Interior, USBR, April 11, 2023, draft Environmental Impact Statement: Colorado River Operations.

The Gila River sub-basin to the Colorado is a different situation even through the drought. The 1964 Supreme Court ruling, which granted Arizona exclusive rights to the Gila River watershed, also altered the management of that watershed. The State applied rules since 1919 on top of the Federally mandated priority requirements. Arizona requires that all new rights established after 1919 also be continually used with no gaps longer than 5 years. It also established criteria giving private water districts an added requirement that water be used within the watershed. ARS 45-101 et seq. Staying within the watershed appears to be a significant and very valuable difference in the current condition of the greater Colorado River watershed and the Gila River sub-watershed.

An example of this approach that has been successful is the Salt River Project, which was established in 1903 by the landowners to be served by it. The Salt River Project is unique in that it was created with the assistance of the federal government as a reclamation project to manage the Salt River watershed for beneficial consumptive use within the community. This facilitated the construction of the Granite Reef Dam, which diverted water into the canal system at the confluence of the Verde River and the Salt River. Did you know that much of the canal routing for SRP was very similar to that of the Hohokam Indians over 500 years ago? Turns out those historic water users were pretty good hydraulic engineers.

Effective management paved the way for the Roosevelt Dam to be funded and constructed. The Bureau of Reclamation built it with a pledge to pay for it given by the lands that would benefit from it. The "on-project" lands were pledged to the project as guarantors that the cost would be repaid. The use of the water for farming made repayment possible. Ultimately, the Salt River Project paid for that dam and constructed seven additional dams along the Verde and Salt Rivers to store and manage the precious water resources of arid

Arizona. Today, they are proposing raising the Bartlett Dam to further enhance water storage.

The lakes behind the dams are not approaching dead pool or empty status, as lakes Mead and Powell are. It should be noted that all "on-project" lands are fed by gravity and within the watershed. The SRP system does not serve lands outside this criterion. I believe that a simple criterion to limit service to lands within the watershed has played a significant role in the state of water in the Salt and Verde River sub-watersheds. SRP has been able to meet a growing and changing water demand despite the 20-year drought the State has been facing.

What can be observed is that the Salt River Basin has served the City of Phoenix and the SRP on-project lands more efficiently than the Bureau of Reclamation has managed the rest of the Colorado River Basin. My assessment places the start of the differences in the original decision not to listen to hydrologists when lands outside of the watershed were allowed to claim rights for the distribution of the river water.

Ok, we might have screwed that one up on the Colorado, so how do we adapt to the current situation and adopt new solutions to our water needs? Is there a way to fix this? I think so, but it will take a lot of collaboration towards a common goal.

Chapter 6: My Example of a Town That Must Adapt – Prescott Valley

Why Prescott Valley

People might ask why I chose Prescott Valley for my dive into a community. There were numerous choices to choose from. What criteria would matter to me? Size? Rate of growth? Water sustainability? Or the towns that struggled with water? Who dried up? Who mismanaged? Hmmm….

Prescott Valley is a medium-sized rural town that is part of a larger multi-government community. It is a growing community that has room to grow sustainably if it chooses to. Prescott Valley is located in the Prescott Active Management Area (AMA), and the aquifer has been deemed not to be in safe yield by the Arizona Department of Water Resources. Current and past local governments have had differing approaches to dealing with the water situation, building up to a struggle. Wells around the edges of the aquifer are experiencing water scarcity issues as the water table slowly declines. How can we better manage this situation for the sake of our children's future?

Figure 8: The Town of Prescott Valley

All that aside, the real reason is that Prescott Valley is where I grew up. Actually, nothing in the Town existed when I was born. It was part of the Rafter 11 Ranch. This is the environment that shaped what my future could look like. I saw the land before the Town existed and have watched the community grow over the last 60-plus years. I worked cattle across the land where the Town Hall now sits. I installed some of those older water services, and I worked with crews installing the main pipelines as the community grew. I was there when the Elephant Eye and Santa Fe wells were tested. They are is still special because they show indications that there could be a deeper, unexplored water source. Now, to see the Jasper well and the possibilities that it opens up as we begin to understand what may be possible. I have watched a sleepy little blue-collar town grow from the dust into so much more.

I understand our roots as a family, and the very similar roots as a community are intertwined as we have grown together. I have worked and watched the beginnings of a Town, a community, and a lifestyle. I have watched the growth through the years from a perspective few can even conceive because they were not there. Through that, I have a unique understanding of what growth in a community means and how water makes that possible.

I have spent a large part of my lifelong career exploring and understanding our water resources in this great state. That has led me to lead teams to produce new water resources through exploration, reclamation, and importation. I have worked on methods to change water's character to meet the varied needs of our varied uses. My firm provided much of the field work to verify the aquifer conditions of the Big Chino Aquifer for the groundwater modelers to create their assessment of the underground supplies. My purpose has been to help communities grow in sustainable ways so they can thrive and support their residents across the whole state. However, my passion lies in where I came from.

I come from a long line of stewards of our resources. When your family lives off the land, and you want to be sustainable as a family or community, you take care of what takes care of you. If not, you can't live there because it won't support you. This water primer is part of my stewardship as I work to educate people about the past and help them innovate for the future. The Town of Prescott Valley is where I have the most passion and knowledge to share. It is also a place that I think might consider being more proactive in securing its long-term water sustainability. I hear stories that sets artificial limits on what could be. With this primer, I offer a strategic path to save the Colorado River and get wet river water into the Prescott AMA for our grandchildren and great-grandchildren to have possibilities for opportunities, just like each of us who have lived here has had.

Sources of Water

Sources of water in Prescott Valley include groundwater, surface water, and reclaimed water supplies today. Groundwater is primarily used to serve the potable needs of the community. Surface water today supports recreation, industry, and jobs in the area. Reclaimed water supplies augment groundwater through recharge and in-lieu uses within the community, such as watering the Stoneridge golf course using effluent instead of groundwater. Continuing to improve on the technologies that allow us to be effective stewards of these resources as we move into the future is necessary so that Prescott Valley may grow sustainably and prosper. When our community prospers, we all have an opportunity to prosper as well.

Groundwater in Prescott Valley is pumped from the Upper Agua Fria Sub-Basin and the Little Chino Sub-Basin. Most of the groundwater development in the Upper Agua Fria Sub-Basin has been found in the middle zone, located above 1000 ft below the

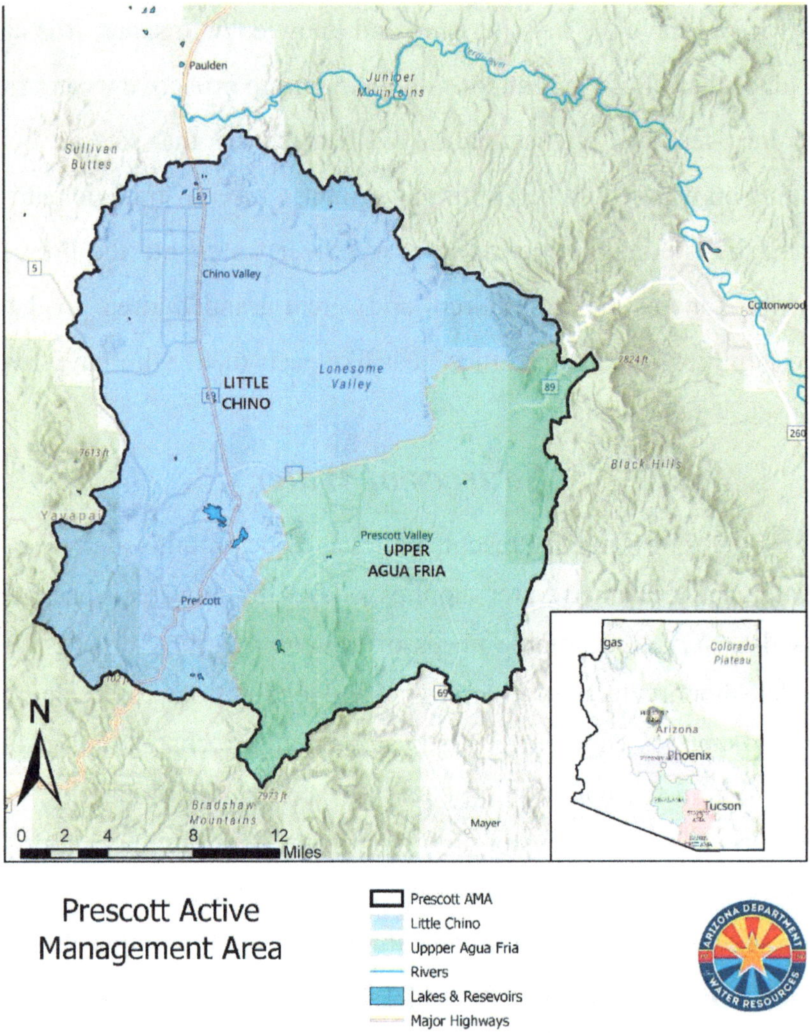

Figure 9: Prescott AMA Groundwater Basins

ground surface and below the prevalent clays that represent the upper zone near the surface. There are some exceptions that appear to be in deeper water that is not fully understood, in my opinion.

Possibilities include deep volcanic sources and/or a connection to surface waters from the Bradshaw Mountains in the basin's groundwater, possibly through fractures connecting to the base of the aquifer. It is believed that Upper Agua Fria is predominantly unconfined with some isolated confined conditions. This indicates how complicated this aquifer actually is. As deeper wells are drilled in promising areas, this understanding will grow, and that will inevitably change perspectives of the local groundwater supplies. It is possible that this new information, once developed, could drastically change the groundwater modeling that ADWR had performed on the AMA. Interesting possibilities in a very complex groundwater system.

The main source of surface water available to Prescott Valley is Lynx Creek. This source is available year-round due to Lynx Lake and Fain Park Lake, which collect Stormwater, store it, and make it available for public recreation throughout the year by retaining the water behind two dams. Today, this water supply augments construction-oriented mining and jobs in the community, as well as agriculture, putting food on our tables.

The future uses envision the portion of this water being beneficially used for potable purposes, supporting 100-year assured water supplies for future homes, using a renewable resource instead of groundwater. In the interim, future uses will include direct use for irrigation and recharge of surface waters that have not yet been used for their highest and best uses. This will mitigate the loss of this water resource from the region. Using this water across the community will improve the Town's sustainability and ensure that the rights to use this water remain available in the community for future homes to sustain growth in the region.

Reclaimed water in Prescott Valley is possible because of the centralized sewer and Class A+ treatment facilities that the State compelled the Town to install in the 1990s to protect groundwater supplies. At that time, the sheer number of septic tanks in the area was threatening the quality of the groundwater supplies with human-caused nitrogen contamination. The only viable solution was a centralized collection and treatment system, allowing the sewage to be properly cleaned so it would not continue to contaminate the drinking water quality aquifer.

Once the Town had the backbone facilities in place, the thinking shifted from just water quality to also include water quantities, and the ability to reuse and recharge treated effluent within the community was actively pursued. It still is today, with more intention than ever before. The Town has spreading basins along the western banks of the normally dry Agua Fria River, as well as recharge efforts in the river bottom of the Agua Fria River as it passes the treatment plant. The Town has been exploring different methods of using wells to get the water into the aquifer. A portion is also reused on parks and the Stone Ridge golf course, which saves groundwater for higher and better uses. This is a great start; however, there is a significant amount of lost water using these recharge methods. A future and more efficient augmentation method would be direct potable reuse. This would likely involve a water treatment plant following the wastewater treatment plant to ensure the quality of water delivered to the public exceeds all drinking water standards.

Imported water does not currently exist in Prescott Valley unless you consider that water in the bottle you are drinking from. However, it has been in the planning phases for years. Why, because the Prescott AMA, including the Upper Agua Fria Sub-Basin, has been deemed out of balance and is said to be mining groundwater. The Big Chino aquifer, located northwest of Chino Valley near the confluence of the headwaters of the Verde and Agua Fria Rivers, has been designated for importation into the region through legislation, ARS 45-555.

This importation of water is the next lowest-cost solution to increasing water supplies for the region. Initially, it will likely be recharged in the Little Chino Aquifer below the community of Chino Valley near the cone of depression caused by groundwater pumping. It is believed that this aquifer is hydraulically connected to the Upper Agua Fria, generally along Highway 89A. As the current cone of depression in the Little Chino is reduced, it is believed that water will again spill into the Upper Agua Fria Aquifer, potentially putting both aquifers on a path towards a state of safe-yield.

Importation from the Big Chino doesn’t happen because we wish it so. The communities within the Prescott AMA must act to secure the water and prepare to import it in the future. Securing the water means purchasing the land that has those rights. Preparing to import it means building infrastructure to withdraw the water from the ground and transport it to the cone of depression. In doing this, the existing wells in the area will be spared as the groundwater surface begins to increase across the region. I believe that this will be the next lowest-cost solution towards developing sustainable water supplies in the Prescott AMA. This effort will have a beneficial long-term impact on every well in the Prescott AMA, regardless of whether they are in one of the legacy Towns and Cities or in the County; everybody will benefit. The sooner the Legacy communities that have the right acquire the water, decide to secure the land having the water rights, the lower the ultimate cost of securing sustainable water resources for the region will be. Once the water rights are secured, methods for extraction, delivery, and recharge into the Little Chino aquifer can legitimately be budgeted, have funding identified, and pursued. The City of Buckeye and the Town of Queen Creek just secured approvals to augment their water supplies through importation. Prescott Valley, Prescott, and Chino Valley need to be more proactive in this area of water resources.

People who are concerned about the potential need to mitigate damage that pumping from the Big Chino aquifer may cause to the headwaters of the Verde River should consider the solutions offered in the last two chapters of this book. A way to accomplish mitigation, if necessary, will be inherent in getting Colorado River water into the Prescott AMA. Based on field studies used to calibrate groundwater modeling, it is not believed that mitigation will be necessary. Those studies that the Salt River Project agreed with set a pumping rate that the scientific methods used to assess the conditions indicate should not impact the headwaters of the Verde River. The addition of the mitigation clause was an acknowledgement that there are still unknowns when estimating conditions below the ground surface that we cannot see.

Saving the Colorado River could allow a significant amount of river water to ultimately be delivered to northern Arizona. This new source of imported water would allow the Prescott AMA to continue to grow into the next century in a sustainable manner. Chapters 8 and 9 discuss how it might be possible to grow sustainably because we can save the Colorado River.

What Water Makes Possible in the Community

Water is needed to build communities, help them grow, and sustain them. The wise use of water is water use that is sustainable to grow and support a community so that it is available for future generations to live a good life without anybody having to leave. That means a community that may not be in safe yield would have to reduce its population until safe yield could be sustained if other solutions could not be developed. Who must leave if no growth is the only option considered? Those that came 1st, those that came last, the oldest, the youngest? Do viable possible solutions exist today, or do we need to be more creative?

Growth allows much of the future costs to be borne by those joining us in this special place. A community that thrives will embrace technological change that can help sustain our future and the future of our children and their children. Some of the evolving industries in Prescott Valley that are adopting and adapting to change include:

Sand and gravel mining has replaced gold mining. It supports the construction industry, which builds our roads, utilities, homes, and businesses. Operations in Lynx Creek rely on renewable resources to help protect and manage our water resources. Without this industry, transportation problems would worsen, as new roads would not be feasible, and older roads would be unable to be repaired. Utilities, roads, and infrastructure would suffer when repairing and upgrading because aggregates and concrete would not be available. Home and business construction would become significantly more expensive or have to stop because materials would no longer be available locally, increasing the associated carbon footprint and potentially adding to global warming concerns, which could reduce the available water even further. Water is vital for using these resources that are vital for sustaining a community.

Solar power is used to generate electricity to support the growing community while reducing water demands related to energy. Many energy sources convert water to steam to turn turbines that provide us with electricity. Photovoltaic solar uses no water in the actual production of electricity, thanks to technology. The "local water" that would be used to generate our electricity will be saved for the environment.

Regional distribution centers, such as Ace Hardware, provide jobs with low water use. These distribution centers reduce the carbon footprint of our goods, indirectly preserving our precious water resources while mitigating global warming.

Service industries such as restaurants, boutiques, and consultants all use water in different ways to support the community and are vital to its growth.

Medical services, including hospitals, medical offices, dental offices, nursing homes, and rehabilitation facilities, rely on water to clean and disinfect the tools used to keep us healthy. Without water, these facilities would become incubators for disease.

Agriculture is evolving to use water more efficiently. Flood irrigation is being replaced with sprinklers and, in some cases, subsurface drip irrigation. Watering schedules are more rigorously managed to mitigate and avoid wasting water. Where possible, renewable surface water will keep our precious groundwater for higher and better uses.

Manufacturing of products. Whether it is aluminum framing, motors, or a foundry, water is required for processes, cleaning, and human needs.

If a community stops growing, there is a significant likelihood that, within a generation, it will begin a painful journey towards ghost town status. Not enough water will stop growth and result in a migration away, taking all those services with them. Those of us who are left will not have the conveniences that we have today because the experts who made those conveniences possible have left the community.

Environmental and Social Water Balance

Environmentally, today, the Prescott AMA, of which Prescott Valley is a part, has been deemed out of balance due to "groundwater mining," where water is withdrawn faster than it is recharged. This means that if nothing changes, someday, the community will run out of groundwater because it is using more than is restored through natural and artificial means today. **Let me repeat that, if nothing changes, someday, the community will run out of groundwater.** So, who will leave so that the population is reduced to a level at which groundwater mining will no longer exist? Who?

No growth does not solve this problem! It makes it worse! Without doing anything to change the current situation, the only solution is to shrink the community until a safe yield on the groundwater supply is achieved. What businesses get banned? What does this do to my quality of life? How long before the community dries up and blows away? Do the newest arrivals have to leave first? Maybe it is the youngest graduates who leave. Maybe the ill are forced out? What happens to the community when these decisions have to be made? How does the community stay vibrant and important?

It was determined long ago that when a community stops growing and begins to shrink, buildings go empty, crime increases, services decline, and the people stop wanting to be there. It happens in big cities and little towns. Can these communities continue to exist? Yes, but they are no longer major contributors to our ability to live a good life. In some cases, they become safe harbors for criminals and illicit activities.

In Arizona, it was determined long ago that the importation of water was going to be necessary. On September 30, 1968, President Lynden B Johnson signed the Colorado River Basin Project Act, enabling the Central Arizona Canal (CAP) project to move forward. Construction began in 1973 and was completed 20 years later. The day that water importation became reality occurred in 1985 when the CAP made its first deliveries to the Harquahala Valley Irrigation District west of the Phoenix Metropolitan area. The system was completed in 1993 when the CAP opened for business all the way to Tucson, Arizona. There have been discussions of taking the CAP further, possibly serving areas in Cochise County if water supplies could become sustainably available.

New water sources are required so that our rapidly growing communities can continue to be places we want to live, allowing us to lead the good life, and leave possibilities for our children and grandchildren. I share this as an example of the time it takes to complete these massive projects. The above only accounts for the time to plan, design, and construct. It fails to address the time it takes the parties involved to agree on who is in control, who is subordinate, and how it will be paid for. Kicking the can down the street, claiming "we can't" for whatever reason, will not result in an acceptable situation. Solutions take a long time to facilitate and it is time to act in the Prescott AMA.

In some parts of the state, the water imported from the greater Colorado River is not currently available. Other solutions have been devised to provide interim solutions to the growing water needs. Prescott and Prescott Valley have a Statutory right to water from the Big Chino aquifer northwest of Chino Valley if they are willing to pay for it. Growth pays for this in ways that can reduce the costs of water so that other community endeavors can be pursued. Water is required for growth, and growth is required to produce:

- Affordable housing for our service support industries. Our nurses, teachers, firefighters, police, construction workers, waitresses, cosmetologists, and other essential workers need an affordable place to live that they can call HOME.
- Quality education for our youth through attracting bright teachers and educators. If they have a place to live affordably, they will come and teach our brightest and best students to be the best they can be.
- Medical and other professionals, when housing is available, and demand is high. Not every medical professional is a well-paid doctor. Most are struggling until well into their

careers. Without water, we cannot have affordable housing, and we need affordable housing, even for professionals.

- Local contractors to help us live our good lives. Without good jobs contactors can't hire good help. Without water and affordable housing, we can't draw the labor to give contractors good help and us quality homes, businesses, hospitals, and roads.
- And so much more.

I have heard it said in the community that no growth doesn't hurt, even if it doesn't help. I strongly disagree; if there is already a shortage, which ADWR currently claims, no growth by itself still makes the shortage worse over time. Growth can and will help fund the solutions that meet our water resource needs well into the future. However, those of us who are here must act to accomplish this so that those who come in the future can help cover the costs of their ability to enjoy this great place with us. It is the same philosophy that each current resident has followed as you moved into the community. I stated before that the Town was part of the Rafter 11 Ranch 60 years ago. Every improvement to the water system over that time has been paid for by growth. It is a proven approach that has allowed everybody who is in the area to be here. It is also the responsibility of those who are here to ensure those who come after us have the same opportunities. We must leave possibilities for opportunities for future generations.

What Needs to Happen in The Prescott AMA and Prescott Valley?

The community needs to continue to adapt; however, maybe a bit faster than they have been. I offer the following, some of which they are already contemplating, while others may be new.

- Continue to pursue effective recharge methods while reducing lost water. Spreading basins and in-stream recharge are operating with some success. Vadose zone wells should be explored in some areas, while injection wells may be required in others.
- Pursue a near-zero loss approach of direct potable reuse in the community. That means authorizing the construction of infrastructure to ensure safe water supplies continue to be delivered.
- Continue implementing conservation measures. The 5th management plan from ADWR is a place to start if you are new to this. Sections cover municipal, industrial, agricultural practices, and more. Another place to find conservation ideas around your home are any of the Universities and Jr. Colleges around the state. They know Arizona and understand conservation in their specific region of the state. I would recommend the U of A and Yavapai College as starting points for the general public.

- Improve stormwater management for the purpose of increasing groundwater recharge and reducing evaporation, transpiration, while reducing physical flow downstream losses that are not controlled by surface water rights. Local surface water rights should be used as necessary to ensure that the adjudication process does not remove the water from the region by court action.
- Improve public education by telling the whole story to a possible ultimate end so the population understands the entire process, thereby reducing the fearmongering. Telling people we are going to run out of water without a solid plan to develop sustainable water is foolish when the research has been either completed or initiated to develop these supplies for future generations. People deserve to know this.
- Pursue legal access to the Big Chino water, allowing the planning and the ultimate importation of said water into the AMA. This buys time for our declining aquifer until we achieve the goal of importing Colorado River water.
- Ultimately, the pursuit of the Northwestern Arizona Aqueduct from Lake Mead would support more sustainable water resources in Arizona. This effort would include the communities of Kingman, Seligman, Ash Fork, Williams, Flagstaff, Chino Valley, Prescott, Prescott Valley, and Dewey-Humboldt, greatly spreading

> the cost among the users. Considering the sustainable growth this could inspire in the area, the benefits would be significant. More people to pay the costs, sustainable water supporting sustainable jobs, housing, and the economy. This aqueduct could connect to the Big Chino infrastructure, making it useful and viable for generations into the future. This also provides a means to mitigate the Big Chino aquifer if it is determined necessary after the fact, meeting the terms of the agreement with the Salt River Project. It could also be used to support an Indian settlement required to use the Big Chino water on an interim basis. Water from the Colorado River would be available to recharge the Big Chino while also recharging the Little Chino and providing resources to the community.

Sometimes I feel guilty when I tell you something without grounding it. Little things like serving northwestern Arizona with water from the Colorado River. The next chapters offer some insight as to how this might happen. It won't be an overnight success. It will take decades to accomplish. This book creates a story of possibilities that could realistically happen, and while we are at it, preserve each of the existing improvements up and down the river, benefiting everybody who uses the recreation, consumes electricity from the grid, or eats the food produced by our farmers and drinks water.

Chapter 7: Wet Water Possibilities for Arizona

How to create long-term sustainability over the Next 100 Years. First, we must recognize that solving the water problems for Arizona is really a much larger issue. Our water situation is tied to six other states and Mexico. Therefore, solving Arizona's water resources will have to include helping the other stakeholders with their water concerns as well, not financially, but physically and mentally. Financially, all who benefit will participate in one form or another. This is my big wet water dream, and I am sure others will have variations to this approach, and that is good. I am introducing this to start that conversation so we can continue to develop sustainable water resources for our great State. I should not say "start the conversation" since work actually began in the late 1800s during our territorial days before statehood. So let this conversation and today's efforts continue the conversations that produce the actions that will allow us to develop sustainable water supplies for the next century and beyond.

What is already being done

Let's start with what we are already doing as a state and water-conscious community in addition to any Federal water requirements:

- Arizona has been working with different water-using sectors of our economy to attempt to develop a comprehensive plan that will help the State get through the declining water availability from the Colorado River. This plan is intended to set a strategy that will allow all economic sectors of our economy to adapt to changing water conditions and work towards improving them. Short-term solutions may include paying entities not to use their water allocation for a period of years, but that is not a long-term fix. It can be a stop-gap measure until we create solutions that do not reduce our food supplies or any other commodities using this water for beneficial purposes.

- Arizona has created some of the most comprehensive groundwater laws in the country because we recognized over 80 years ago that we were using more than Mother Nature was able to replenish. Not everybody agrees with how the ensuing laws were promulgated because they impacted property rights. However, a significant amount of good has come out of these laws to protect the ability of Arizona to grow responsibly. Those laws have created plans that have been updated approximately every 5 years since their inception to reflect new knowledge about our groundwater systems and our attempts to attain a safe yield for each groundwater basin in Arizona, not just the AMAs and INAs.
- Our codes have been adopted in a manner that promotes water conservation in our homes, businesses, and communities. Promoting low-water-use fixtures.

Pretreatment regulations for industry. Standards for cleaning sewage for reuse and recharge to directly and indirectly augment existing water supplies.

- We have improved our farming, agricultural, and mining practices to reduce waste and water pollution from them. We have grown to view the products these activities produce as required to sustain human life, as we have grown to appreciate them. The efficiency of water use has significantly improved due to the methods employed, and further improvements are expected as businesses adopt the new methods being developed. It makes economic sense to continue to improve our efficiencies around our water use. Business embraces that kind of rational, maybe for different reasons, but with the common outcome of helping to make our water resources more sustainable.
- Barlet Dam is scheduled to be expanded to capture more water from the Verde River watershed within the State. This will increase the amount of water available from surface waters within the state. The current proposal is to double the volume of water that could be held in the reservoir in an attempt to increase our water sustainability by further reducing the water lost during periods of wet weather.
- Arizona has been a leader in exploring means and methods to augment groundwater supplies. These include recharge of unused surface waters and reclaimed waters from both

industrial and municipal sources. Arizona has stored excess Colorado River water using recharge methods in groundwater basins along the Central Arizona Canal (CAP) and in each of the initial AMAs in an attempt to reach safe yield conditions while also storing water for the future. Much of the water stored along the CAP was excess water along the main stem of the river that Arizona won access to in the 1964 Supreme Court ruling. It was being stored for times such as these when no excess water is available. It can be recovered from the underground storage and delivered to the demand. Is this enough to solve our scarcity concerns? No, but it helps, as does every contributory effort towards developing more sustainable water supplies.

- Arizona is currently developing rules and infrastructure to take treated sewage and make it potable for direct service to the population. This effort will eliminate the incidental losses that occur when we reuse indirectly and recharge aquifers. We have the ability to do this safely and on a large scale. It is part of our efforts toward building sustainable water supplies as we eliminate past water loses. This could stretch our potable supplies up to 10% in some areas of the state through reduction of losses using other methods.
- Arizona has evaluated and designated specific aquifers, through statute, that are located adjacent to but outside designated AMAs that can be used to pump groundwater

from outside AMAs for uses inside those AMAs. The Phoenix AMA has already secured two importation efforts for the City of Buckeye and the Town of Queen Creek. Other communities should consider securing their imported water before it is a crisis.

As of the writing of this book, there are five separate basins where water transfers have been legalized for intra-state water relocation. Intra-State Water Transfers is the relocation of water from areas with low use and high capacity to Aquifer Management Areas (AMAs) throughout the State. These serve as a temporary stopgap until more sustainable resources can be imported, such as interstate and international solutions. These are plentiful aquifers that have available demand or have been historically augmented by surface water recharge and have been identified by Statute for use in augmenting supplies in Aquifer Management Areas. The following summarizes each of the five basins where transfers are authorized by Statute:

- ARS 45-552 Transportation of groundwater withdrawn in the McMullen Valley basin to an active management area

  My interpretation of this law is that it allows water to be pumped from the McMullen Valley if the land was purchased prior to 1988 for the purpose of transporting groundwater. This is only available to a city, town, private water company, groundwater replenishment district, or the Arizona Water Banking Authority. The information found indicates that the City of Phoenix is the primary beneficiary of this Statute due to land and possibly district ownership prior to 1988.

The water could be removed via wells and put in a dedicated pipeline to the existing Central Arizona Canal (CAP) near Vicksburg Junction. Approximately 5 to 8 miles of pipe could be laid, with much of the route along the existing Hwy 60, to mitigate new environmental studies and damage, and reduce National Environmental Policy Act (NEPA) financial and time impact on the project.

Figure 10: McMullen Valley Basin

- ARS 45-553 Transportation of groundwater withdrawn in the Butler Valley groundwater basin to an initial active management area

My interpretation of this law is that any of the initial AMAs, Phoenix, Pinal, Tucson, or Prescott, have the legal right to withdraw water from this aquifer for importation into those AMAs.

The water could be removed via wells and put directly into the CAP to serve Phoenix, Pinal, and Tucson. Serving Prescott would likely require some sort of exchange of water rights in northern Arizona with a holder of those rights in southern Arizona. Salt River Project is a potential partner but others could also develop.

The Central Arizona Groundwater Replenishment District (CAGRD) is a special taxing district that is likely the primary beneficiary of this water. ARS 48-3171- 3180 created the CAGRD to allow some users to meet the requirements of a 100-year Assured Water Supply. It was formed to acquire, store, and recharge water on behalf of its members, who are typically made up of subdivisions and public and private water providers inside the AMAs. It was developed as a workaround for subdivisions proceeding in provider areas that do not have designations of an assured water supply. Members pump

Figure 11: Butler Valley Basin

groundwater, and that is replaced elsewhere in the AMA using a renewable supply such as surface water or reclaimed effluent.

- ARS 45-554 Transportation of groundwater withdrawn in the Harquahala irrigation non-expansion area to one of the initial active management areas.

  My interpretation of this law is that it allows the withdrawal and transport of groundwater from this basin to an initial AMA or La Paz County, with limitations on the depth and rate of withdrawals and requirements for studies, monitoring, and reporting to ensure compliance with per-acre caps.

  Water is again removed through wells and pumped 3 to 8 miles to the CAP canal. The route would follow existing roadways to mitigate the need for environmental studies and expedite the ability to permit and construct improvements. As of July 18, 2025, the City of Buckeye and the Town of Queen Creek within the Phoenix AMA have both secured the right to transport water from this aquifer to help in sustaining their water resources for their communities. Other communities in the Phoenix, Pinal, and Tucson AMAs are also eligible.

Figure 12: Harquahala INA

- ARS 45-555 Transportation of groundwater withdrawn in the Big Chino sub-basin of the Verde River groundwater basin to the initial active management area.

My interpretation of this law is that it allows the withdrawal and transport of groundwater from this basin to any city or town in an adjacent initial active management area. The annual allotment is grounded in the historic irrigated use of 3 acre-ft/acre, with withdrawals capped over annual and decade periods. This potential can increase to 8,068 acre-ft per year if the CAP water rights are relinquished and tribal water rights are settled. As of this book's writing, the tribal settlement has been executed, although full implementation has not yet been completed. The settlement with the Yavapai-Apache Nation is entitled to 4,610 ac-ft/yr from the Cragin Reservoir on the Mogollon Rim. In 2024, Senator Mark Kelly and Senator Kyrsten Sinema introduced a bill to ratify and fund the Yavapai-Apache Nation's water rights. The funds still need to be appropriated and implemented in full.

Obtaining the wet water from the Big Chino will require wells and approximately 30 miles of conveyance infrastructure to reach the cone of depression in the Little Chino sub-basin, where recharge could likely occur. This has yet to be funded. When constructed, recharge would then potentially be able to reduce the scarcity of water

Figure 13: The Big Chino

experienced at the edges of both the Little Chino and Upper Agua Fria sub-basins that make up the Prescott AMA. It will slow and possibly mitigate the groundwater mining for today, buying time for bigger, more permanent solutions, such as importing Colorado River water once we save the Colorado.

- ARS 45-555.01 Transportation of groundwater withdrawn in the upper San Pedro groundwater basin, Douglas active management area

  My interpretation of this law is that if a well owner was providing groundwater to the Douglas AMA prior to December 1, 2022, they can continue to do so if they hold a certificate of convenience and necessity in amounts that do not exceed peak historic uses for the 10 years prior.

  The Arizona Water Company – Bisbee is the user we could confirm was benefiting from this water resource to meet Bisbee water demands within the utilities' certified area of service. It is possible that other communities are benefiting from this water but it could not be confirmed.

Arizona will continue to search for new methods and means to augment and protect its water supplies; however, this may still be insufficient to meet all the future needs of the state as it continues to grow over the next 100 years. To offset that possibility, importing water from the ocean may be our next best move.

What could be next

The first step in what is next is for communities in the AMAs to begin to secure the water they need to import in the short term. This will buy time for the more sustainable solutions, which will take decades to complete, to be developed.

- Prescott and Prescott Valley in the Prescott AMA have the framework in which they could work together to purchase land in the Big Chino, allowing them to begin development of the infrastructure to serve the communities. In this situation, wells must be drilled; a conveyance system of either pressure or gravity must be designed and constructed; a receiving facility, such as an above-ground treatment or aquifer storage facility, must be developed.
- The towns, cities, designated water providers, along with the CAGRD, who have rights in the McMullen and Buttler Valleys, could work together to develop a means to convey the water into the needed communities.
- Cities and towns in the Phoenix AMA are currently working to secure water from the Harquahala basin. This water could be transported through the existing CAP system if excess capacity is available. The CAP delivery system currently delivers water to or generally near many of the communities that could use the water and

greatly reduces the overall cost of importation. With less water available from the Colorado river, (in the short term) there is likely capacity in the canal system to convey these waters. As the river is restored, additional engineered storage along the canal system will help increase overall capacity to meet these transport needs.

The concept of importing water from the Pacific Ocean and the Sea of Cortez has been discussed for years. With treatment capabilities improving to make this not only possible but more cost-effective, the possibilities for this solution to become a reality are increased to the point of almost no doubt. I say almost no doubt it will happen, but due to all the moving parts, it will not be easy or fast.

Ultimately, this plan directly impacts California and Mexico, and we must work with both over the course of the coming years to accomplish sustainability for Arizona and to help both Mexico and California sustain their situations as well. Indirectly, it will benefit each of the 7 western states and Mexico using Colorado River water by reducing our reliance on this limited supply. There is a possibility that solutions implemented for Arizona could also greatly benefit the upper basin states. If this becomes a reality, those states could also participate in the cost of the infrastructure to facilitate oceanic importation. In exchange, water reallocations could be accomplished using current data and infrastructure improvements that help us all.

The following bullet points outline my thoughts on a plan that can be realistically accomplished, yielding significant benefits for each participant. It is centered on the importation of external water resources: the use of ocean water resources to augment local supplies.

- Build a desalination plant off the coast of California near Los Angeles.
    - One 500,000 ac-ft/year plant to serve the Metropolitan Water District of Southern California, serving both Los Angeles and San Diego, allowing that amount of water to be retained in the river for others. This water could then be held in Lake Mead and Lake Powell for use in the northern communities and tribal lands of Arizona once the lakes reach safe operating levels. In the short term, this source would also go a long way in keeping these lakes from reaching dead pool status unable to generate electricity we all use.
    - A second 500,000 ac-ft/year plant to serve Arizona. This could use the existing multi-barrel pipeline aqueduct alignment from Parker Dam to LA, reverse its operations and fill Lake Havasu from the Pacific

Ocean. A significant amount of reengineering of the aqueduct would be required but the National Environmental Protection Act (NEPA) requirements could be greatly mitigated cutting significant project costs and time delays.

- This would add 500,000 ac-ft/year of water to the Central Arizona Canal system. Initially, to replenish much of the water reduced in Arizona due to the drought, and secondly, for the future growth of Arizona. This water supply would likely benefit those along the CAP canal since Lake Havasu is also the headwater of the CAP system. Taking this approach with the existing aqueduct would significantly reduce, if not eliminate, most of the NEPA requirements, saving time and money.
- These combined projects would add 1,000,000 ac-ft/yr to water supplies to the river system. 500,000 ac-ft/yr would be derived by serving California coastal cities with desalinated seawater, allowing commensurate river flows to remain in the river. The second 500,000 ac-ft/yr of treated Pacific Ocean would be conveyed back to the Colorado River adding to the capacity at Lake Havasu. Consider it new water to the watershed. The opposite of taking water out of the watershed.

Figure 14: Colorado River Aqueduct

- Build a desalination plant in Mexico sourcing water from the Gulf of California (Sea of Cortez).

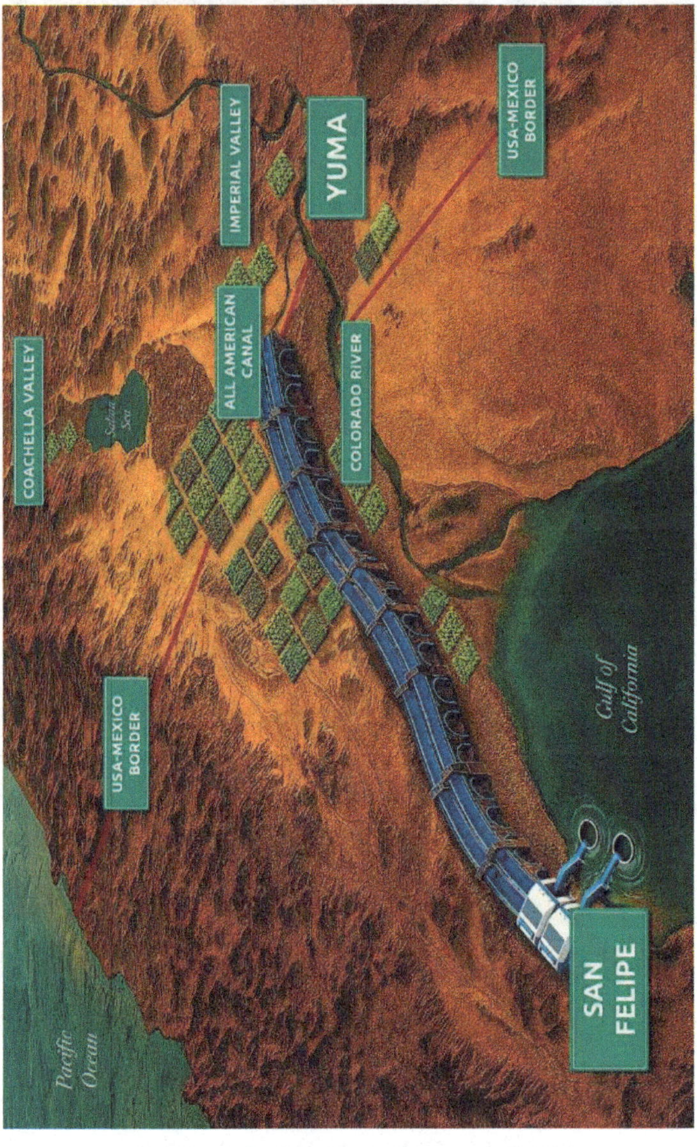

Figure 15: Sea of Cortez Aqueduct

- A partnership with Mexico to replenish the salad bowls of the Colorado River and ensure the long-term growth of farming must be afforded. This would be treated seawater from the Gulf. The alignment could follow existing highways to mitigate environmental concerns for the pipeline construction. This pipeline would likely terminate at the All-American Canal but could also be extend up to the Parker Dam, serving Lake Havasu, and adding potential capacity to the CAP if necessary. However, if an exchange agreement with the out-of-watershed California farmers could occur, up to approximately 3,100,000 acre-feet per year could be retained in the river system by providing a secure and reliable supply of water to the All-American Canal for their use. That is a lot of water. It accounts for more than Arizona's entire current main stem river allocation. It would allow for a reorganization of the water rights in both the upper and lower basin. By itself, this solution could contribute mightily to the saving of the Colorado River!!
- Water, treated to Colorado River quality could be distributed along the entire route to serve our growing farming needs in the southwest US and Mexico. The potential savings to the river could be several million acre-feet of water, depending on how big we make the

infrastructure and how much of California's and possibly Mexico's irrigation rights are replaced by this source. The point is that there is water in the river to save by adding new sustainable resources to the overall water system. An engineered solution. Mexico could be made whole and get the entire 1.5 million ac-ft/year that the US committed to back in 1944, while we also reduce our own farming demands on the river water supply.

Keeping more water in the river will also benefit our power supplies. Each of the dams in the lower basin produces hydroelectric power that serves the western grid. When the lakes behind these dams drop, the amount of power that they are capable of producing declines. Operating these lakes with higher volumes of storage increases the hydro-electric power potential serving cities such as Phoenix, Las Vegas, Los Angeles, and San Diego and everybody in between who is served by the electric grid.

These two large water projects could combine to greatly increase the total water available today along the Lower Colorado River even more if California agriculture participates. As presented, Arizona could see as much as 4.1 million ac-ft/yr added to our current allocation without consideration for the water Mexico might choose to leave upstream if they choose to replace it with seawater. One of

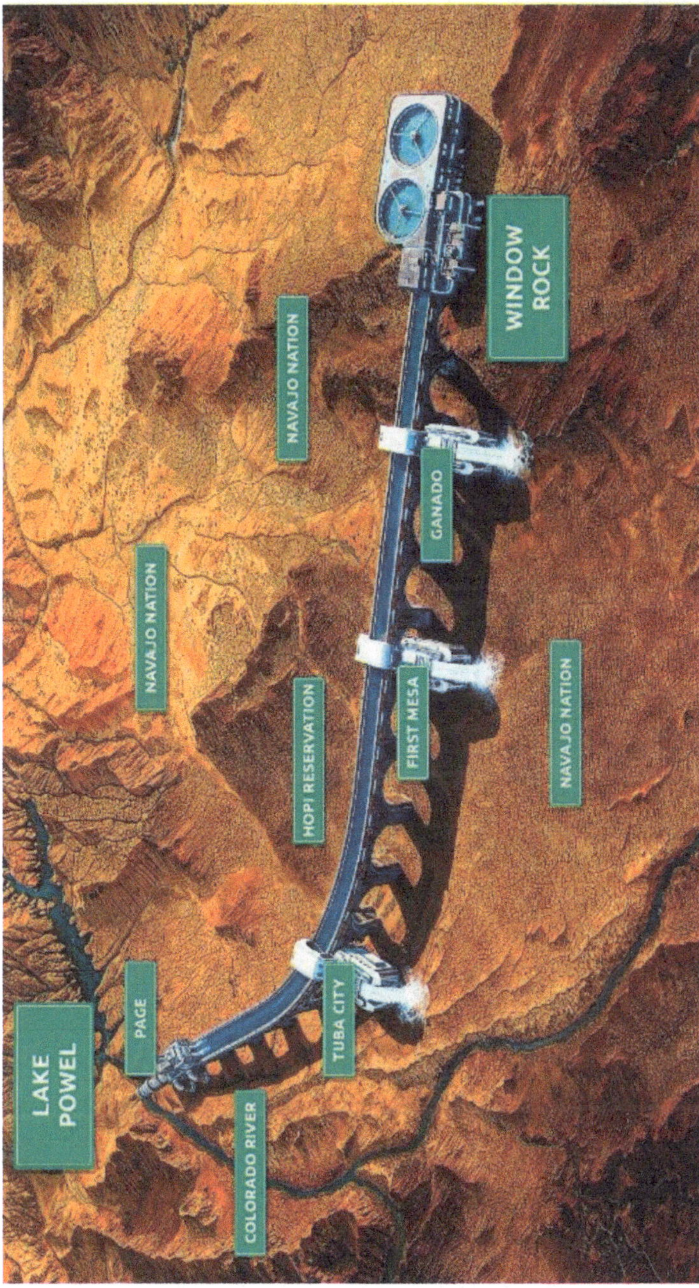

Figure 16: Nava-Hopi Aqueduct

these projects, the Nava-Hopi Aqueduct, could source water from Lake Powell for deliveries to Tuba City, First Mesa, Ganado, and Window Rock. Over time, this could be extended to also serve Cameron and Kayenta on the Navajo Nation and Holbrook, Show Low, and Payson south of the Little Colorado River.

A second project could be the development of the Northwestern Arizona Aqueduct. This could serve Kingman, Seligman, Ash Fork, Williams, and Flagstaff on one route while a second branch leaves Seligman through the Big Chino and connects to the existing Big Chino pipeline serving Chino Valley, Prescott Valley, Prescott, and Dewey-Humboldt. This conduit would also serve as a means for mitigation of the Big Chino should that be deemed necessary in the future. Mitigation would be the direct recharge of surface water into the Big Chino aquifer, should past exportation by Prescott or Prescott Valley of groundwater from the Big Chino negatively impact the flows of the Verde River.

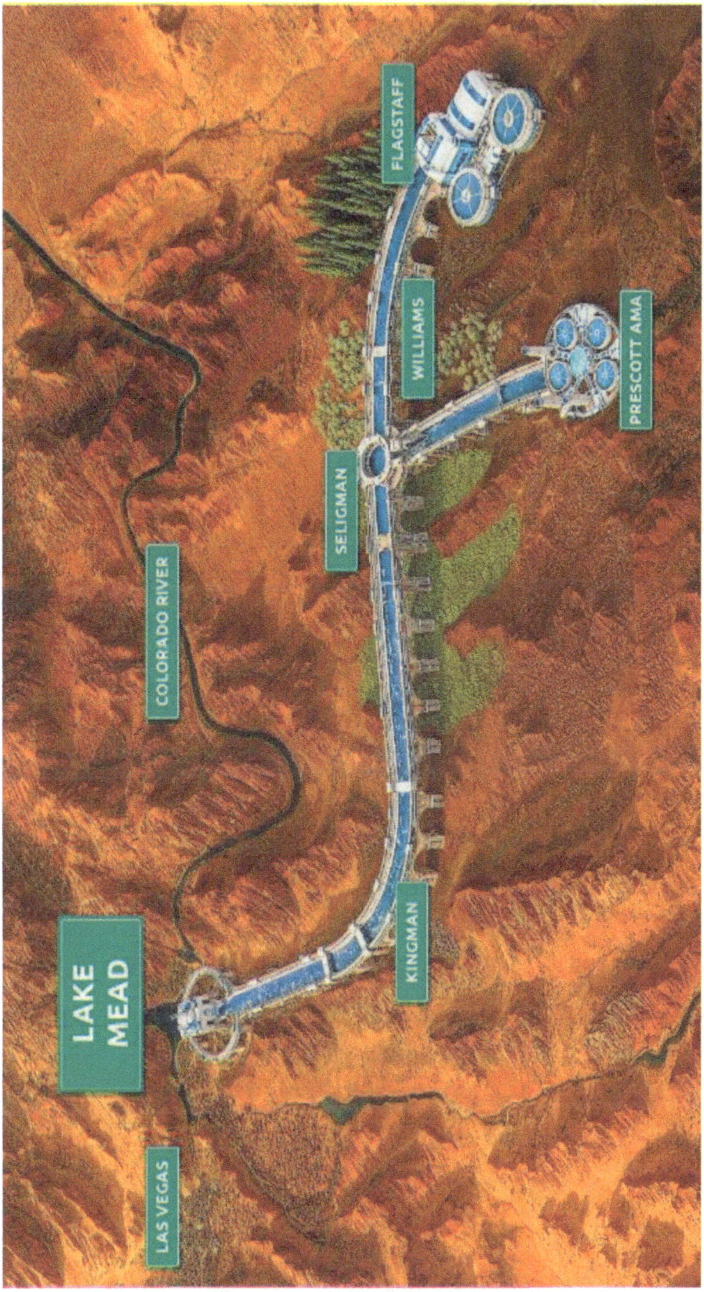

Figure 17: Northwest Arizona Aqueduct

The actual distribution and quantities used can vary, but the concept is the same. Importation of water to reduce demands on the river today, while allowing for added uses tomorrow. This not only allows northern Arizona to flourish with imported renewable water resources, but it also allows the Federal Government to partially meet its obligations to our tribes across the state, including the Navajo and Hopi reservations.

An important side benefit of this approach is the increased sustainability of the energy produced from both the Glen Canyon Dam at Lake Powell and the Hoover Dam at Lake Mead. These two hydroelectric power producers impact much of the southwestern and western US. With this in mind, these users could also be part of the solution as they, too, will benefit. This benefit will help spread out costs across many industries, several states, and upwards of 50 million users plus by the time this gets implemented.

Many people will claim that energy is the reason that desalination and pumping projects of this magnitude will not be possible. However, with advances in nuclear fission and nuclear fusion, these plants could have their own power plants since a cooling water source will not be an issue. Once nuclear fusion advances further, it will produce fewer undesirable byproducts in the waste stream. This approach makes these types of systems self-sufficient. They will not contribute to the grid overload. They will continue to produce water when the grid goes down.

Chapter 8: Possible Ways to Proceed Towards Water Security & Sustainability

What is a Public-Private Partnership (PPP)?

A typical Public-Private Partnership (PPP) involves public government agencies and private parties, including businesses of many kinds, collaborating to jointly plan, design, implement, finance, construct, and operate infrastructure for the public benefit. In most cases, it is the government that identifies a project that it assesses would benefit from private participation to enhance efficiency, innovation, funding, or all three. In some cases, the project is identified by the private sector and brought to the public sector for them to bring it forward.

In most cases, the first step is a feasibility study to determine if the PPP is, in fact, the best approach. This study will assess the environmental, financial, technical, and social needs of the project. This is often concluded with project structuring, defining standards, outcomes, and methods to achieve the outcomes. When our government does this, it typically involves several years of in-depth evaluation. Getting these projects started often requires a significant amount of effort because regulators tend to get in their own way, trying to ensure that every consideration is addressed satisfactorily for each stakeholder. In reality, that concept rarely results in a satisfactory solution for everybody. Somebody will almost always have grounds to dislike any solution.

At this point, the private sector participation can be solicited if it is not already a part of the team. This is where performance measures of the project are defined. Who bears what risks between the public and private sectors? This often leads to defining who is responsible for financing, construction, operations, and management of the project.

Once the Public-Private Partnership has been defined, the work and fun can begin. The private sector often designs the infrastructure, while the public sector ensures it conforms to laws, rules, and codes while meeting the overarching intention. Who finances the work often defines who manages, operates, and owns the project. A government entity will typically manage a publicly financed project. It is normally paid for through tax dollars, utility assessments, and any other fees the Government can find ways to justify. Alternatively, a privately financed project will typically be managed by the private sector, with government oversight from a distance. This approach may be financed through banks or venture capitalists and will be paid for with user fees for a period of time, similar to a toll road, but for a pipeline or canal transporting water, it is called a "wheeling fee". This fee is intended to cover operations, maintenance, and replacement costs of the infrastructure.

Ideally, a PPP will deliver a project for public use that takes advantage of each party's strengths to deliver a superior project for a reasonable cost in a shorter timeframe. That can mean higher-quality products delivered over a shorter period of time at a lower cost to the public than conventional public sector project development. One of many solutions that is often selected results in lower taxes while only charging the actual users for the cost of the infrastructure. If you are a customer of the water utility that is augmenting its supplies with imported water, the whole customer base will pay a similar rate. However, new users will also pay an impact fee to offset the costs to existing customers. Impact fees are intended to pay for capital costs, while a customer's bill typically pays operations costs. Ideally, somewhere in those costs is a recovery or replacement cost, because everything wears out eventually, even if it takes 50 years to do so.

There are numerous ways in which Public-Private Partnerships can be structured and paid for. Each situation will offer unique circumstances that will shape how the PPP works so that it is as fair as possible for all users.

Likely partnerships should include the parties involved, as they will ultimately be paying for the infrastructure, similar to the on-project landowners within the Salt River Project, which delivers both water and power.

Water Innovation

Water has shaped our civilizations, industries, and our daily lives throughout much of human existence. Records of harnessing water for the greater good date back over 3000 years before Christ. The Hohokam Indians harnessed water in New Mexico and Arizona 500 to 700 years ago. This is still evident in the canal system that is operated by the Salt River Project (SRP) in the Phoenix metropolitan area today. SRP followed a similar canal routing to that of the Hohokam because it worked and was efficient.

The Indians of that era were pretty good hydraulic engineers. They understood how to work with what nature gave them. They built canal systems that sprawled across this desert to irrigate their crops, to support their families and communities. What they were missing was all the innovation that had happened since they disappeared from these deserts. Mankind's curiosity didn't stop with the Hohokam's creativity. It is this curiosity and creativity that make it possible to overcome drought today in ways the Indians 500 years ago could not accomplish.

Initial mechanization using water included using the power of water to turn equipment, so that tasks such as grinding grains could be done more efficiently. This required the equipment to be operated to be located near a natural water source that contained sufficient energy to turn a water wheel, thereby completing tasks such as grinding grains.

The invention of the steam engine in the early 1700s brought forth the ability to move water from one place to another in quantities and pressures that were not possible before. These engines, connected to pumps in combination with advanced canals and pipelines, resulted in significant economic expansion. The Erie Canal is an example of facilitating waterborne freight, where it could not exist before, just as the Central Arizona Project canal facilitated renewable water supplies, where there were limited supplies before.

As the 1900s dawned, the emphasis moved towards the quality and treatment of waterborne disease. This enhanced public health and urban development. Chlorination and filtration techniques created early in the century contributed to the reshaping of how we lived. These techniques allowed larger populations to move from rural country homes, with miles between them, to urban developments, with neighbors within a few feet of each other. Advances in treatment allow urban neighbors to still have safe water and wastewater systems that may not have been possible in past centuries.

More recently, innovation has allowed us to refine our exploration, treatment, and distribution capabilities so that we can be better stewards of this precious resource. Academic and private research efforts have redefined our abilities to treat and monitor our water and the systems that deliver it. This has been a catalyst in our efforts to optimize our water resources so that they are available for future generations.

Water innovation culminates in the need for responsible water stewardship. History has shown that when governments, industry, and communities communicate and collaborate, great advances have happened in short times for the benefit of the greater populations.

Today, we have the technology and human know-how to use the tools to treat ocean water to drinking water standards if necessary. Membrane technology today is very advanced and very capable of handling this task.

My Opinion About How to Proceed

Everybody working to find new and more efficient ways to use our water must continue. Who knows what great idea somebody might have that could save the water we use. Those little things add up when done collectively. Repairing leaks, efficient yard watering techniques, low flow fixtures, tankless water heaters, rooftop solar, and more will all help to conserve our water supplies. **Individually, we can each contribute to the long-term objective of water sustainability without water scarcity.**

At a much larger scale, we need to take care of business in a timely manner to meet our near-term water needs and our distant needs. Obtaining water from the Pacific Ocean or the Sea of Cortez will likely take decades to complete. Therefore, infrastructure within the state, which involves moving water from designated aquifers to meet current needs, must be planned and constructed. At a minimum this would augment the water demands in the populated areas of the Prescott AMA, Phoenix AMA, Pinal AMA, and the Tucson AMA. These projects also take time also in part because regulatory reviews can be protracted when we are doing something new. Due to the time required to plan, design, permit, and construct, each of the communities within the legacy AMAs should budget for and pursue these water sources.

While the short-term needs are being met, concurrent planning and initial permitting of the larger, more sustainable seawater solutions must also begin. (Unless a better idea develops.) Due to environmental and economic concerns, the locations for treatment facilities and the routes for new conveyance infrastructure will take a significant amount of time to complete. Using existing infrastructure and following existing rights-of-way will significantly help these concerns, but not alleviate them.

Working to secure additional water from the Colorado River for the newer AMAs could be possible if the Colorado River could be saved. These include the Santa Cruz, Douglas and Wilcox AMAs. This could justify the expansion of the CAP to our communities southwest of Tucson.

The Least Cost per Ac-ft Today

I believe the most cost-effective first step could be removing the Metropolitan Water District of Southern California, serving Los Angeles and San Diego, from the Colorado River system by constructing a desalination plant capable of producing 500,000 acre-feet per year. That would be done in exchange for their high priority water right to the water in the river. The District would have a more reliable source that would not be subject to the weather cycles of the river. It also would not be subject to future cuts should the drought continue. Once the treatment plant is complete, it will immediately trigger a 500,000 acre-feet per year saving to the river that could be used in Arizona. This would add to Arizona water rights and help to mitigate the impact cuts Arizona has taken due to the drought on the river and our low priority rights.

River quality water would be delivered to the current treatment facilities that treat the river water, or a separate facility could be constructed by the District at the desalination plant. The parties that would pay for this would depend on how the project was structured between public and private entities. However, the cost would be spread to all who benefit from the improvements. The District would have greater reliability, and Arizona would have more river water and a higher priority right for a portion of that water.

Next, a second desalination plant could be constructed near the first to convey 500,000 ac-ft/yr of river quality water back to Lake Havasu at the Parker Dam. This would also require reequipping the existing aqueduct so that water could be pumped back to the river instead of from the river as it currently operates. Taking this approach eliminates the need to perform extensive environmental studies on the pipeline route as it currently exists and would be rehabilitated to operate in reverse. Reducing the environmental studies alone could save decades in design and 100s of millions of dollars in the process between the costs of studies and the cost impacts to all of us for delays that could be avoided. This doesn't eliminate NEPA, just greatly streamlines it with the efficiency of taking advantage of past efforts.

The net result of these two combined efforts would put 1,000,000 ac-ft of water back into the system. Half by replacing the District's water right and assigning it to Arizona, and the other half would be adding new water put into the system. I would envision that this new water would have a very high priority, provided the treatment systems used to create this resource remain operational. This would add sustainability to a significant portion of Arizona's current allocations.

During periods of high river flows, these facilities could be turned down to only produce what is needed at the time once the river is restored to its grandour. This would keep the cost of the water lower over time. During periods of drought, the treatment facilities could make up for the water shortages on the river to mitigate the need to reduce allocated water to Arizona. Operational flexibility will exist if we design it into the system.

The Biggest Potential Impact on the River

The greatest impact from a water availability perspective could potentially come from the Sea of Cortez. This solution could provide the greatest benefit to the river, Mexico, and Arizona, while giving the farming communities a more consistent and reliable water supply. It will likely have the highest cost as well, not because of distance, but because of volume. That translates into serving more people to share the overall cost of the project with.

The river could see up to 3,100,000 acre-feet per year, which is currently conveyed out of the watershed to California farmers. This solution would retain that water within the watershed. Those California farmers would trade reliability and consistency of water quality for their priority rights on the Colorado River. Arizona could get some of the highest priority rights in the river in exchange for helping to develop this new resource. This solution has the potential not only to help Arizona but also the entire Colorado River system, including both upper and lower basins, as well as Mexico.

Seawater would be treated to river-quality levels and delivered to farms in Mexico and California. Mexico could increase the lands irrigated in its part of the farm belt, increasing the number of jobs and improving lifestyles. This could also be a combination of multi-barrel pipelines and open channel lined ditches with outlets along the way to the All-American Canal. The All-American Canal, which currently conveys Colorado River water to the California farmers, would be repurposed to convey treated seawater. The farming community should have a more consistent water quality than previously provided by the river, potentially reducing their costs to grow our products. This source would not be subject to Mother Nature and fits of drought along the river.

Northern Arizona and the upper basin of the Colorado River Watershed could also benefit from retaining this water in the river system. Reservoirs throughout the watershed could retain more water and possibly be increased in size to accommodate more storage, all because of a relatively short pipeline when compared to the CAP canal, which used a very successful model to build and pay for the infrastructure.

Building infrastructure to treat 3,100,000 ac-ft/yr is a significant undertaking of treatment, pumping, pipelines, and canals, but it is not unrealistic when we consider we have already built the All-American Canal to do just that. The CAP canal into central Arizona conveys approximately 2,200,000 ac-ft/yr. So we know it is possible and that we are capable of doing this kind of infrastructure for the benefit of everybody. Keeping the route as short as possible for the largest benefit is important to keep costs under control. The California farmers have the largest high-priority rights to the Colorado River's water. The All-American Canal is capable of conveying that water to the farmers. Treating an additional 1,500,000 acre-feet per year will also restore water for Mexican farmers. The potential savings to the river with both countries participating are 4.6 million ac-ft/yr. This would greatly improve water levels in Lakes Mead and Powell, restoring capacity and making more water available throughout the entire Colorado River Watershed. This will indirectly improve our food and electricity supplies. That in turn, will help to make lives better for our children and grandchildren.

The potential combined impact of these two major projects could put the river system into a safe yield situation while still accommodating growth within the watershed. These two projects would also greatly reduce or eliminate the amount of water that is shipped outside of the lower basin portion of the natural watershed. The net result would be a significant increase in the overall sustainability of the Colorado River. It may even curtail the river from dying while allowing growth to pay for it. All combined, the potential net savings and additions to the river could amount to as much as 5.6 million ac-ft/yr.

500,000 ac-ft/yr	Exchanged with the Metropolitan Water District of Southern California
500,000 ac-ft/yr	New desalinated water added to the watershed by reversing the Colorado River Aqueduct
3,100,000 ac-ft/yr	Exchanged with the farmers in Coachella and Imperial Valleys
1,500,000 ac-ft/yr	Replacing Mexico's water allocation with their agreement in exchange for their allocation
5,600,000 ac-ft/yr	**Potential water that could be restored to the river**

Think about this. This river is overallocated during droughts such as the one we have been experiencing. The total allocations to the river due to government regulation have been 16.5 million ac-ft/yr. Of that, 7.5 million ac-ft/yr was allocated to the lower basin of which Arizona is the largest land mass of any state contributing to the watershed. Include Mexico's 1.5 million ac-ft/yr and the lower basin total becomes 9.0 million ac-ft/yr. If we can reduce that demand by 5.6 million ac-ft/yr the river might fully heal and allow us to responsibly grow further into it. We could thoughtfully reallocate the lower basin water in a manner that is sustainable for the river system despite climate change and drought conditions. Arizona could continue to grow into its future while saving the Colorado River by creating engineered sustainability on the main stem.

Additional benefits to improving the sustainability of our water supplies in the Southwest United States and Northwest Mexico are found in our food and energy supplies. The fruits and vegetables produced in the region would have a sustainable ocean supply and would not be subject to the drought conditions we see today. This will benefit people across North America. Keeping water in Lakes Mead and Powell will enable those dams to continue producing sustainable hydroelectric power, thereby keeping the Greater Los Angeles, Las Vegas, Phoenix, and everyone in between fully energized with a green energy supply. This benefit helps produce much-needed electricity using a renewable, non-polluting resource that we get to use again downstream, creating more electricity, water for our plants and livestock, as well as providing drinking water for ourselves.

The benefits are far-reaching, helping people who may have never seen the river or its grandeur when food and energy are considered. The benefits affect not just the US but Mexico as well. This is a possible scenario where international neighbors can collaborate to develop long-term, sustainable solutions addressing our water, food, and the needs of our people and communities in both countries.

Chapter 9: A Strategy to Allow Arizona to Grow While Saving the Colorado River

It will likely take decades to get one or both of the desalination projects on line and delivering water to current Colorado River users that are outside of the watershed; therefore, interim strategic steps need to be taken to allow Arizona to continue to grow. These steps will buy our water resources time while we build the next major infrastructure component toward the goal to bring long-term water sustainability, without scarcity, to the region. Growth will then pay for the lion's share of the cost of these improvements to the sustainability of our water supplies for the benefit of all of us. Interim steps involve using our local water supplies as efficiently and effectively as possible. The ultimate steps involve importing water from the Pacific Ocean and the Sea of Cortez which will make our supplies much more sustainable for centuries into the future. This strategy focuses on the rights to use the water and physical water deliveries for our use. Regulatory and legislative oversight and evolution will also occur is one reason for the amount of time that this effort will take.

Two parallel paths should initially be pursued simultaneously. One to facilitate the intra-state importation between groundwater basins. This allows the State to grow thoughtfully for the coming decades. The other path is to facilitate importing water from both the Pacific Ocean and the Sea of Cortez. This needs to be a serious discussion as soon as possible, because getting past the political and environmental concerns will take a significant amount of time. Nothing will get constructed until these concerns are satisfied for all interstate and international parties to" *Saving the Colorado River"*.

Conservation and State-Wide Groundwater Management

Conservation efforts throughout the state should continue. Where appropriate, rainwater harvesting and stormwater management to facilitate less groundwater reliance while simultaneously augmenting groundwater supplies should be initiated if possible and continued if it exists. We must continue to use water prudently and for its highest and best uses. As a public, we can do the little things to conserve around our home and work. It doesn't mean you must limit your shower to 60 seconds, but using a low-flow shower fixture would be nice. Cumulatively, that will continue to make a difference. It is a part of the reason Arizona has done as well as it has in managing our water supplies so that we all still have wet water.

I never thought I would say this because I still believe that the State took property rights with the 1980 groundwater law without proper compensation. However, that ship has sailed, and I have had to adapt. Now I look at the regulatory agency and recognize that its efforts have been generally good, but they need the legislation to allow them to go further in protecting our water supplies outside of AMAs as well as inside them. Those advances need to continue as we work to bridge the gap between now and importation; intra-state, interstate, and international importation.

Intra-State Importation

Step 1: Legacy AMA Towns and Cities, including the CAGRD, must secure the water rights of groundwater basins and storage sites that have been designated by legislation. These water resources will provide the interim water needed until the Colorado River can be restored. Without securing the right, the ability to obtain the water could be lost in the future due to either business or political actions.

Once the river is restored, these basins could also be restored as we use excess water from the Colorado to recharge the water borrowed. Having control of the physical water also allows communities to begin to plan for the infrastructure required to transfer the wet water into their communities. As of this writing, I congratulate the City of Buckeye and the Town of Queen Creek for having taken this first step. They can now legitimately start planning infrastructure needs.

Step 2: Plan for the infrastructure required to import the various water resources to the communities owning the rights. Planning includes what the project looks like and how it will be paid for. This will define whether it is completely government-owned and operated or includes a public-private partnership.

Step 3: Construct and operate the infrastructure. In some cases, it may be possible to use the CAP to transport the water through a wheeling agreement. Others, such as those in the Prescott AMA, will have to construct the entirety of the infrastructure required, because the existing CAP canal does not serve them. In all cases, wells and some pipelines or canals will be required.

These efforts buy much-needed time to implement solutions that are truly sustainable.

Seawater Importation

Step 1: Begin and, in some cases, continue initial discussions with water management counterparts in California and Mexico. These are two separate negotiations that would ideally occur simultaneously. Discussions with the State of California will include the coastal cities using Colorado River water from Los Angeles to San Diego, as well as the Metropolitan Water District that serves them. Add to this the farmers of Imperial and Coachella Valleys, as they become major players in its success. Discussions with Mexico will include the cities of Tijuana and Mexicali, along with farmers along the Colorado River Delta, led by the Mexican government and its appointed agencies. To top it all off, our own government will have interested parties from several agencies involved, including the Bureau of Reclamation and the Environmental Protection Agency.

Because there will be many different representatives from both public and private entities that will need to be involved with these transactions, it will take a significant period of time to facilitate agreements that everybody can accept. Based on the time it took for Arizona and California to settle disagreements over the Colorado River, this could take decades to resolve. We humans usually try to get a little extra for ourselves. It is in our nature to do so, but it often costs us time and money along the way.

Step 2: Once the political, regulatory, and environmental concerns have been addressed, we can begin to plan the details needed to finance, construct, and operate the infrastructure. This effort will be Herculean and will take years to complete, and will cross both interstate and international lines. The answers to the questions generated will likely produce the best, lowest-cost solutions that will serve each community in reliable and sustainable ways.

This effort will answer these questions and more:

- Do Tijuana and Mexicali get served from the United States or through a facility built in Mexico?
- Does the desalination plant off the Los Angeles Coast get constructed to serve both California and Arizona?
- Can the Colorado River Aqueduct be successfully reversed to serve Arizona?
- How big will the desalination plant in the Sea of Cortez be planned at and in how many phases?
- What is the quality of water that each of these desalination plants must produce?

- Can existing road right-of-ways in both the US and Mexico serve as the routes required to transport the volumes of water being considered, or will alternative routes be more advantageous?
- How much water will be delivered to Mexican farmers before getting to the All-American Canal?
- How much are the California farmers willing to cooperate with this effort to save the river and our mutually connected lifestyles?
- How much are Arizona farmers and Arizona tribal communities willing to cooperate with this effort?
- Which importation project will occur first, California or Mexico?
- How does each part of these projects get financed and paid for?
- Does infrastructure from Mexico stop at the All-American Canal, or does it continue to Parker Dam, where it can serve the Central Arizona Project?
- Who will benefit from the new water created by importation, and what will the water rights associated with this water look like?

Step 3: With some of the questions answered from step 2, we can begin to design and construct the actual infrastructure required to treat and transport the water. I am curious which of the desalination paths will happen first. Who has the greater need and desire to accomplish this will be the ultimate answer to that question.

Infrastructure for the California facilities will include a desalination plant, an independent power supply such as a small nuclear generating station, and waste salt handling facilities to avoid creating environmental concerns in the ocean. This will allow responsible treatment of the seawater to either drinking water standards or Colorado River water quality without polluting the ocean or stressing the power grid. It is likely that drinking water standards will be met initially, and blending will be used to bring costs down for river quality water deliveries. Drinking water standards for the California cities and Colorado River quality for the volumes being pumped back to the river. This effort will also include the installation of new pumps and rehabilitation to reverse the flow in the Colorado River Aqueduct, allowing it to deliver water to Lake Havasu instead of taking water out of the Lake.

The result of this effort could look like this:

Potable Desalination Plant	Up to 500,000 ac-ft/yr	*Benefit to the River,* 500,000 ac-ft/yr left in the main stem. *Benefit to Arizona,* 500,000 ac-ft/yr left in the main stem for use with high priority rights. *Benefit to California,* greater sustainability and reliability.
River Quality Desalination Plant	Up to 500,000 ac-ft/yr	*Benefit to the River,* 500,000 ac-ft/yr of new water. *Benefit to Arizona,* 500,000 ac-ft/yr of new water rights.

This has the potential to add 1,000,000 ac-ft/yr to the lower main stem of the river, with all of those water rights potentially belonging to Arizona. The 500,000 ac-ft/yr that was traded for desalinated seawater could be used anywhere along the main stem because it is water generated by the river system. One possible use is that it could greatly help both Lakes Mead and Powell avoid dead pool conditions, keeping power being generated for millions of users. In the end the highest need at the time will define how the water will be first used. The 500,000 ac-ft/yr that is newly added water will be available for use anywhere from and including Lake Havasu to all uses along the CAP canal or the lower reaches of the main stem below Parker Dam. This source would add to the potential distribution capabilities from Lake Havasu. It could mitigate the effects of temporary water losses due to drought while also creating a new source of water and water rights along the Colorado River.

Infrastructure for Mexican facilities will include a desalination plant, power, and waste salt facilities, similar to what is required in California to treat seawater to river water quality for delivery to farming communities between the Colorado River Delta and to American farmers served off the All-American Canal. It will require a new aqueduct, canal system, or combination to transport the treated water north to the end uses. Some would say this could source from San Felipe, while others might say it might source from Puerto Penasco. That choice should include the Mexican government and its impacted citizens, as well as costs, when considering all situations. What is proposed in this book is just one idea.

One significant advantage of this project is that it could add up to 4,600,000 ac-ft/yr of water that could be saved and reallocated up the entire main stem of the river while helping deliver more water to Mexico and their communities along the river. Mexican farmers,

Towns, and Cities along the lower Colorado River Basin will potentially have new opportunities develop while the existing opportunities expand because sustainable water will be available. By providing an alternative source of water for existing uses outside the river watershed, more water generated within the watershed can be used to shore up shortages and facilitate responsible growth along the main stem of the river. This is akin to making new water because we are using desalinated seawater as our new resource that has not fully been exploited, but has untold volumes of water that we can clean up for our use.

Climate scientists indicate that the seas will be rising and salinity will be dropping over time due to glacial melting caused by climate change. Both of those conditions benefit this potential solution. It is possible that salts produced by desalination of seawater could be repurposed to benefit sea life that might be struggling with declining salinities as glaciers melt and the ocean chemistry changes.

The following table describes my hypothetical approach as to how the water might be redistributed as a result of the desalination facilities. This effort could look like this:

Potable Desalination Plant	To meet potable needs above the current 1,500,000 ac-ft/yr allocation	*Benefits Mexican Towns and Cities* along the main stem of the river.
River Quality Desalination Plant	Up to 1,500,000 ac-ft/yr plus any new needs.	*Benefits Mexican agricultural communities* along the main stem of the river.
	Up to 3,100,000 ac-ft/yr	*Benefits California farmers* in the Imperial and Coachella Valleys.
Up to 3,100,000 ac-ft/yr of Redistributed River Water (This could be 4,600,000 ac-ft/yr if Mexico agrees to help)	Undefined quantity to meet the needs of Lake Mead and northwestern Arizona through a future aqueduct	*Water is restoring Lake Mead.* Eventual aqueduct from Lake Mead to Kingman, Ash Fork, Chino Valley, Prescott, Prescott Valley, Williams and Flagstaff.

Up to 3,100,000 ac-ft/yr of Redistributed River Water (This could be 4,600,000 ac-ft/yr if Mexico agrees to help)	Undefined quantity to meet the needs of Lake Powell and northeastern Arizona through a future aqueduct	*Water is restoring Lake Powell.* Eventual aqueduct from Lake Powell to Tuba City, Cameron, First Mesa, Ganado and Window rock while serving the Navajo and Hopi Reservations.
Reallocation of the Main Stem Water	Undefined reduction in the overall allocation of the river to reflect long term averages	Impacts all of the Upper and Lower Basin States and Mexico along the Colorado River.

With this strategy, Arizona could have sustainable water resources for well beyond the next 100 years. The Colorado River could be saved from over-allocation and overuse while allowing all the communities within the watershed to grow responsibly. People who live in the arid southwestern United States and northwestern Mexico will not be forced to migrate away in search of water. Through the development of more sustainable water supplies, we improve our food and energy supplies as well as our ability to build our technological tools, of which most require water somewhere in their manufacture, and many will help us achieve our goal of mitigating water scarcity through achieving water sustainability. These are just a few of the ways that water is vital for us to all live a good life in the place we choose to be.

Financing

People will ask, "How do you propose to pay for this?" The answer is the same as before: the end users and beneficiaries will pay for this. Growth spreads that cost around, but each and every one of us will benefit, and we will each pay a portion of the cost as we use the power, drink the water, and eat the food that these solutions allow. Projects such as these are often government-financed, with the end users paying the cost over time. Some entities will be the face of that effort, just as the farmers were the guarantors for the CAP canal, but we are all paying for that through how we purchase what the farming makes possible. This includes our meats, vegetables, and textiles that we consume. A twist to some of these projects may include private financing, but the end users still pay the cost over time.

Alternatively, if we do nothing, the water runs dry. When that happens, we lose the power, lose the water, and have less food. It gets more expensive to live a less satisfying life. People will move away. Services will diminish, once thriving communities become stagnant, dreary places to be, with little to offer people who do stay.

Those of us who think these solutions are grandiose and ludicrous, I offer some of the major infrastructure projects our water supply financed the construction of in just the lower basin of the Colorado River:

- Glen Canyon Dam created Lake Powell for water storage, flood control, recreation, and power generation. It was built to store water and control releases under the 1922 Colorado River Compact, which was not fully ratified until the 1964 Supreme Court settlement. It is one of the two largest man-made lakes in the United States.
- Hoover Dam created Lake Mead for water storage, flood control, recreation, and power generation. It too was built to store water control releases under the 1922 Colorado River Compact. It is also one of the two largest man-made lakes in the United States.
- Davis Dam created Lake Mohave for water storage, recreation, flood control, power generation, and to meet the delivery requirements for Mexico's river allocation. It was a part of the 1944 treaty with Mexico, giving them an allocation of 1,500,000 ac-ft/yr;
- Parker Dam created Lake Havasu for water storage and recreation, enabling the Colorado River Aqueduct serving Los Angeles and San Diego, as well as the Central Arizona Project serving Phoenix, Tucson, the Gila River Indian

Community, and farming communities throughout the region. The dam also produces power generation.

- Imperial Dam created the Imperial Reservoir for diversion into the All-American Canal, serving California farmers, putting food on North American plates. This dam also serves to reduce the silt load on the river before sending Mexico its water allocation within the terms of the 1944 treaty.

In each of these projects, the water storage benefited farming, towns, and cities as water supplies were engineered to be readily available all year, and not just seasonally, as determined by the natural water cycle of the river. One significant benefit of engineered storage behind tall dams is the ability to generate large quantities of electricity, greatly adding to the number of people who benefit from these projects. Finally, recreation is made possible where it could not have existed before in the large man-made lakes that formed behind the dams. The projects presented in this book to save the Colorado River will also save each of these dams from becoming obsolete, incapable of producing the power you need to charge our cell phones, electric vehicles, and lifestyles.

When we consider the number of people, businesses, and communities that the above projects provided a positive impact on, we realize that the cost of these projects ultimately gets paid for by many people over a long period of time. We can find an affordable way to pay for this because the cost will be spread far and wide. Each of the beneficiaries of the above-listed infrastructure projects will also benefit from these projects to different degrees, as will new potential users in northern Arizona who do not currently have access to Colorado River water.

If there is one thing that could greatly reduce costs and the time to get things done, it is simplifying the environmental hoops and hurdles that must be pursued to accomplish this. One big step could cut decades out of this effort by reducing the bureaucratic red tape and overlap of agencies that review projects. The National Environmental Policy Act (NEPA), enacted in 1969, requires federal agencies to assess the environmental effects of their proposed actions before making decisions. Typically, for projects such as these, an Environmental Assessment (EA) and/or an Environmental Impact Statement (EIS) is required to accomplish this. An EA can take a year or more for a project, while an EIS will take several years before peer reviews even begin. Permitting should be a matter of months, not years. This can be fixed.

If we really want to reduce costs, we use existing routes where possible and limit the work to the shorter EA and streamline the review process. Streamline all the Federal and State reviews. We streamline the process, reduce the time taken, reduce the overall costs significantly, and produce wet water faster without sacrificing the NEPA purpose to protect environmentally sensitive situations, whether historic, biological, or geological.

The real questions that remain….

Do we want to solve our water scarcity concerns?

and

Do we have the will to get it done?

About the Author

I (Norm) have been married to my beautiful bride, Karen, for over 40 years. We have two adult children and, as of the writing of this book, two wonderfully curious grandchildren who light up the lives of both Norm and Karen.

I am a licensed engineer in Arizona. My practice has been focused on engineering the water needs of the municipal water cycle. I have been a practicing civil/sanitary engineer in Arizona for over 40 years, and wrote this primer to share some of the knowledge acquired over this time. Any mistakes in this book are all mine.

I was raised on a ranch in what is now Prescott Valley, Arizona, where I learned the value of taking care of what takes care of you. I grew up understanding that if you want to create something, you need to take care of what supports that creation. For me, that meant stewardship of the land and the resources it had to offer.

Interest in water started at a young age when, as a 9th grader at Prescott Jr. High, where I studied biology in water and around my family's corrals for a science project to be entered into the Northern Arizona Regional Science Fair. That project described the microorganisms found in and around family corrals at a very basic level and won the Grand Prize at the Fair that year. Many thanks to my biology teacher, Mr. Jerry Munderlow, for his guidance and help in starting this career path. He was one of the most influential teachers I had because he made learning difficult things fun, and that made learning difficult things much easier to do. May his legacy live on through those of us who had an opportunity to learn from him.

During my K through 12 school years, my learning was not limited to scholastic endeavors. It included learning about the land and its many resources, even though I didn't realize what I was learning at the time. Spending time with my father and grandfather learning about livestock, grazing, water supplies, hydraulics, and geology proved valuable throughout my entire career. My education included ranching and the stewardship of the land, and evolved into a role at Shamrock Water Company, where I learned about serving water to the public.

Growing up working on the Rafter 11 Ranch and the Shamrock Water Company became my outside classrooms, learning how to take care of what takes care of me.

My academic education continued at Northern Arizona University with a degree in Civil Engineering Technology, commonly referred to as Construction Engineering today. This plus experience allowed me to obtain my license in Civil Engineering as well as a General Engineering Contractor's license.

After college, I spent a few years working for Southwest Salt, which later became Morton Salt, in Glendale, Arizona, where I learned about hydraulic mining and salt processing. That experience offered a unique perspective on water and wastewater in Arizona. It translated very well into the water and wastewater fields I found myself emersed in. It gave me a unique perspective that I could offer while I worked for several consultants, learning the water-related consulting business before starting my own firm, Fluid Solutions, in 1998.

Fluid Solutions is focused on the municipal water cycle. Our vision is

The Invention of Solutions for Your Unique Water Resource Needs through Engineering Sustainability into our Water Resources

This means something different for each client we help, and we strive to understand what is important in your unique water situation so we can offer the highest quality help possible.

The pictures below show Norm in 1975, studying microbiology in the corrals on the Rafter 11 Ranch. The other picture is 50 years later, in 2025, as a professional consulting engineer.

1975 9th Grade Jr. High School Student Studying Biology in Water

2025 Professional Engineer Providing Water & Wastewater Engineering Services for over 40 years

Picture of painting by Don Merrifield,1975

Given to my father and then passed to me.

Made in the USA
Coppell, TX
26 July 2026

85558433R00089